Studies in Systems, Decision and Control

Volume 67

About this Series

The series "Studies in Systems, Decision and Control" (SSDC) covers both new developments and advances, as well as the state of the art, in the various areas of broadly perceived systems, decision making and control- quickly, up to date and with a high quality. The intent is to cover the theory, applications, and perspectives on the state of the art and future developments relevant to systems, decision making, control, complex processes and related areas, as embedded in the fields of engineering, computer science, physics, economics, social and life sciences, as well as the paradigms and methodologies behind them. The series contains monographs, textbooks, lecture notes and edited volumes in systems, decision making and control spanning the areas of Cyber-Physical Systems, Autonomous Systems, Sensor Networks, Control Systems, Energy Systems, Automotive Systems, Biological Systems, Vehicular Networking and Connected Vehicles, Aerospace Systems, Automation, Manufacturing, Smart Grids, Nonlinear Systems, Power Systems, Robotics, Social Systems, Economic Systems and other. Of particular value to both the contributors and the readership are the short publication timeframe and the world-wide distribution and exposure which enable both a wide and rapid dissemination of research output.

More information about this series at http://www.springer.com/series/13304

Cheng-ke Zhang · Huai-nian Zhu
Hai-ying Zhou · Ning Bin

Non-cooperative Stochastic Differential Game Theory of Generalized Markov Jump Linear Systems

Springer

Cheng-ke Zhang
School of Economics and Commerce
Guangdong University of Technology
Guangzhou
China

Hai-ying Zhou
School of Management
Guangdong University of Technology
Guangzhou
China

Huai-nian Zhu
School of Economics and Commerce
Guangdong University of Technology
Guangzhou
China

Ning Bin
School of Management
Guangdong University of Technology
Guangzhou
China

ISSN 2198-4182 ISSN 2198-4190 (electronic)
Studies in Systems, Decision and Control
ISBN 978-3-319-40586-5 ISBN 978-3-319-40587-2 (eBook)
DOI 10.1007/978-3-319-40587-2

Library of Congress Control Number: 2016945139

Printed on acid-free paper

This Springer imprint is published by Springer Nature
The registered company is Springer International Publishing AG Switzerland

Preface

Differential game refers to a kind of problem related to the modeling and analysis of conflict in the context of a dynamical system. More specifically, a state variable or variables evolved over time according to differential equations. It is a mathematical tool for solving the bilateral or multilateral problems in dynamic continuous conflicts, competition, or cooperation, which has been widely applied in the fields of military, industrial control, aeronautics and astronautics, environmental protection, marine fishing, economic management and the market competition, finance, insurance, etc.

This book is focused on the generalized Markov jump linear systems which is widely used in engineering and social science, using dynamic programming method and the Riccati equation method to study the dynamic non-cooperative differential game problems and its related applications. This book includes the following studies: the stochastic differential game of continuous-time and discrete-time Markov jump linear systems; the stochastic differential game of linear stochastic differential game of generalized Markov jump systems; the stochastic H_2/H_∞ robust control of generalized Markov jump systems; and the risk control of portfolio selection, European option pricing strategy, and the optimal investment problem of insurance companies. In addition, this book created a variety of mathematical game models to derive the explicit expression of equilibrium strategies, to enrich the theory of equilibrium analysis of dynamic non-cooperative differential game of generalized Markov jump systems. It is to analyze and solve the robust control problems of generalized Markov jump systems based on the game theory. The applications of these new theories and methods in finance and insurance fields were presented.

The main content is divided into the following six sections:

1. The introduction and basic knowledge

This section introduces the basic models and the latest research of generalized Markov jump systems, the research content of differential game theory of generalized Markov jump systems, and the related concepts of differential game theory.

2. The stochastic differential game of continuous-time Markov jump linear systems

From the perspective of stochastic LQ problem, this section studied the stochastic optimal control problem of continuous-time Markov jump linear systems, and then to extend study on the two-person Nash stochastic differential game problem, finally to explore the two person Stackelberg stochastic differential game problem, and to achieve the equilibrium solutions of various problems.

3. The stochastic differential game of discrete-time Markov jump linear systems

From the perspective of stochastic LQ problem, this section studied the stochastic optimal control problem of discrete-time Markov jump linear systems, and then to extend study on the two person Nash stochastic differential game problem, finally to explore the two person Stackelberg stochastic differential game problem, and to achieve the equilibrium solutions of various problems.

4. The stochastic differential game of generalized Markov jump linear systems

This part is to establish the following models: two person zero-sum stochastic differential game, two person nonzero-sum game, Nash game, Stackelberg game, to achieve the equilibrium solutions, and to obtain the explicit expressions of the equilibrium strategies.

5. The stochastic H_2/H_∞ control of generalized Markov jump linear systems

Based on Nash game and Stackelberg game, this part is to establish the Markov jump linear systems models, the stochastic H_2/H_∞ control of generalized Markov jump linear systems models, to achieve the mathematical expression of the optimal robust control.

6. The stochastic differential game of generalized Markov jump linear systems in the applications in the fields of finance and insurance

This part is to establish differential game models of the minimal risk control of portfolio selection, option pricing strategy, and the optimal investment of insurance companies. And regarding the probability measurements of the economic environment as a player, regarding the investors as another player, the differential game models are to achieve the optimal control equilibrium strategies by solving two person differential game problems.

The research achievements of this book are sponsored by two foundations: the National Natural Science Foundation of China, which is named “Non-cooperative stochastic differential game theory of generalized Markov jump linear systems and its application in the field of finance and insurance” (71171061); and the Natural Science Foundation of Guangdong Province, which is named “Non-cooperative stochastic differential game theory of generalized Markov jump linear systems and its application in the field of economics” (S2011010000473). All achievements of this research are counting on the assistances and supports of National Nature Science Foundation of China and the Natural Science Foundation of Guangdong Province. Thanks a lot!

A group of members contribute to the accomplishment of this book, which includes the following: Dr. leader Zhang Cheng-ke, who is the professor; the doctoral student supervisor; the dean of School of Economics and Commerce, Guangdong University of Technology; the executive director of Chinese Game Theory and Experimental Economics Association; the executive director of National College Management of Economics Department Cooperative Association; vice chairman of Systems Engineering Society of Guangdong Province; Dr. Zhu Huai-nian, who is the lecturer of School of Economics and Commerce, Guangdong University of Technology; Dr. Bin Ning, who is the lecturer of School of Management, Guangdong University of Technology; and Dr. Zhou Hai-ying, who works in Students' Affairs Division, Guangdong University of Technology. Team members play a team spirit; have close cooperation; work in unity and cooperation; publish a number of papers, which has laid a good foundation for the completion of this book. The achievements of this book presented in front of readers are the collaborative efforts and hard work of all members of the research group!

Thanks to Zhang Chengke's graduate students Cao Ming, Zhu Ying! They have made a lot of work in terms of manuscript input, format correction, and check the formula, etc.

Special thanks for the help and supports of Guo Kaizhong, who is the professor of Guangdong University of Technology; and Cao Bingyuan, who is the professor of Guangzhou University! Owing to their constant encouragements make this book completed and presented to the readers as soon as possible.

Counting on the References to the scholars quoted in the book, which make the fruitful base of our work!

Although we have made a lot of efforts for the completion of this book, due to the limited level, there must be a lot of shortcomings and deficiencies. Please to criticize and correct.

Guangzhou, China

Cheng-ke Zhang
Huai-nian Zhu
Hai-ying Zhou
Ning Bin

Contents

Representation of Symbol

M'	The transpose of any matrix or vector M;
$M > 0$	The symmetric matrix M is positive definite;
$\mathbb{R}^n$	The n-dimensional Euclidean space;
$\mathbb{R}^{n\times m}$	the set of all $n \times m$ matrices;
$\mathcal{S}^n$	the set of all $n \times n$ symmetric matrices;
$\mathcal{S}^n_+$	The subset of all non-negative definite matrices of $\mathcal{S}^n$;
$\mathcal{S}^n_l$	$\underbrace{\mathcal{S}^n \times \cdots \times \mathcal{S}^n}_{l}$;
$(\mathcal{S}^n_+)^l$	$\underbrace{\mathcal{S}^n_+ \times \cdots \times \mathcal{S}^n_+}_{l}$;
$\mathcal{M}^l_{n,m}$	space of all $A = (A(1), A(2), \cdots, A(l))$ with $A(i)$ being $n \times m$ matrix, $i = 1, 2, \cdots, l$;
$\mathcal{M}^l_n$	$\mathcal{M}^l_{n,n}$;
χ_A	The indicator function of a set A;
$C(0, T; \mathbb{R}^{n\times m})$	The set of continuous functions $\phi : [0, T] \to \mathbb{R}^{n\times m}$;
$L^p(0, T; \mathbb{R}^{n\times m})$	the set of functions $\phi : [0, T] \to \mathbb{R}^{n\times m}$ such that $\int_0^T \|\phi(t)\|^p dt < \infty (p \in [1, \infty))$;
$L^\infty(0, T; \mathbb{R}^{n\times m})$	the set of essentially bounded measurable functions $\phi : [0, T] \to \mathbb{R}^{n\times m}$;
$C^1(0, T; \mathcal{S}^n_l)$	the set of continuously differential functions $\phi : [0, T] \to \mathcal{S}^n_l$.

Content Introduction

This book systematically studied the stochastic non-cooperative differential game theory of generalized linear Markov jump systems and its application in the field of finance and insurance. First, this book was an in-depth research of the continuous-time and discrete-time linear quadratic stochastic differential game, in order to establish a relatively complete framework of dynamic non-cooperative differential game theory. And using the method of dynamic programming principle and Riccati equation, this book derive into all kinds of existence conditions and calculating method of the equilibrium strategies of dynamic non-cooperative differential game. Then, based on the game theory method, this book studied the corresponding robust control problem, especially the existence condition and design method of the optimal robust control strategy. Finally, this book discussed the theoretical results and its applications in the risk control, option pricing, and the optimal investment problem in the field of finance and insurance, enriching the achievements of differential game research.

This book can be used as a reference book for graduate students majored in economic management, science and engineering of universities in learning non-cooperative differential games, and also for engineering technical personnel and economic management cadres.

Chapter 1
Introduction

1.1 Research and Development Status of Generalized Markov Jump Linear System Theory

1.1.1 Basic Model of Generalized Markov Jump Linear Systems

The research of switched systems is mainly carried out with the research of hybrid systems [1–5]. A hybrid system is a dynamic system that exhibits both continuous and discrete dynamic behavior–a system, such as manufacturing systems, weather forecast systems, power systems, biological systems, as well as option pricing models in financial engineering, insurance surplus distribution models, multi-sector fixed asset dynamic input-output models, etc., that can both flow (described by a differential equation) and jump (described by a state machine or automaton). In the process of its operation, a hybrid system often suffers from a sudden change in the environment, internal connection changes between each subsystem in a large system, changes of nonlinear objects, damages of the system components and random mutations, such as human intervention. These phenomena can be seen as a response of the system driven by a class of random events. In general, the state of such a system is defined by the values of the continuous variables and a discrete mode. The state changes either continuously, according to a flow condition, or discretely according to a control graph. Continuous flow is permitted as long as so-called invariants hold, while discrete transitions can occur as soon as given jump conditions are satisfied. Discrete transitions may be associated with events. Such systems are often called hybrid systems in control theory.

When the discrete event of hybrid systems is characterized by discrete switching signals, such important systems are called jump systems. This kind of systems can be described by finite subsystems or dynamic models, and at the same time there is a switch law, which makes the switching between various subsystems.

C.-k. Zhang et al., *Non-cooperative Stochastic Differential Game Theory of Generalized Markov Jump Linear Systems*, Studies in Systems, Decision and Control 67, DOI 10.1007/978-3-319-40587-2_1

A stochastic jump system can usually be described by the following state equations:

$$\begin{cases} \dot{x}(t) = f(t, x(t), u(t), r(t)), \\ r(t) = \varphi(t, x(t), r(t^-), u(t)). \end{cases} \quad (1.1.1)$$

where $x(t) \in \mathbb{R}^n$ is a continuous variable, $u(t) \in \mathbb{R}^m$ is an external signal of continuous control input or continuous dynamic systems, $r(t)$ is a piece-wise constant function valued in a finite set $\Xi = \{1, \cdots, l\}$, usually referred as "switch signals", or "switching strategy" of the system. $r(t^-)$ indicates that $r(t)$ is a piece-wise constant right hand continuous function. When $r(t)$ takes different values, the system (1.1.1) corresponds to different subsystems. $f(\cdot,\cdot,\cdot,\cdot)$ reflects continuous state variables changes of the system, $\varphi(\cdot,\cdot,\cdot,\cdot)$ is the transition function of discrete states, which reflects dynamic changes of logic strategies or discrete events of systems. Obviously, when the switching strategy $r(t) \in \Xi = \{1\}$, the random jump system is degraded as a simple stochastic system. So, a simple random system is a special case of the stochastic jump systems (1.1.1).

A generalized stochastic jump system is usually described by the following state equations:

$$\begin{cases} E\dot{x}(t) = f(t, x(t), r(t), u(t)), \\ r(t) = \varphi(t, x(t), r(t^-), u(t)), \end{cases} \quad (1.1.2)$$

where $E \in \mathbb{R}^{n \times n}$ is a known singular matrix with $0 < \text{rank}(E) = k \le n$, $x(t) \in \mathbb{R}^n$, $u(t) \in \mathbb{R}^m$, $r(t)$, $r(t^-)$, $f(\cdot,\cdot,\cdot,\cdot)$, $\varphi(\cdot,\cdot,\cdot,\cdot)$ are the same as system (1.1.1).

This book is focused on a kind of special jump systems with Markov switching parameters, which is known as Markov jump systems. In such systems, the switching rules determine which corresponding subsystem the system would be switched to at each moment, and the state of the system would be switched to the corresponding state at the corresponding moment. But during the process of the system switching from one mode to another mode, there is no switching rule to obeying, and the switching process between different modes is random. This kind of random switching accords with some certain statistical properties—the transformation among various regime of the discrete event finite set of the system is a Markov jump process, therefore, it can be also regarded as a special case of stochastic systems, called stochastic Markov switching systems (also known as stochastic Markov jump systems, or stochastic Markov modulation systems).

A Markov jump system is constructed by two parts. One part of the system is the state of the system, and the other part is the system mode, which depends on the Markov process, deciding the execution of the subsystem at a certain moment, in order to control and coordinate the normal operation of the whole system.

(1) Mathematical Model of Continuous Generalized Markov Jump Systems

The continuous generalized stochastic Markov jump linear system is described as:

$$E\dot{x}(t) = A(r(t))x(t) + B(r(t))u(t), \tag{1.1.3}$$

where $E \in \mathbb{R}^{n\times n}$, $x \in \mathbb{R}^n$, $u(t) \in \mathbb{R}^m$ are the same as system (1.1.1), and the "switch signals" or "switching strategy" of the system $r(t) \in \Xi = \{1, \cdots, l\}$ is a Markov chain with finite state. Ξ is the state space. Define $\Pi = [\pi_{ij}]_{l\times l}$ as the transition matrix of Markov process $r(t)$, and the transition probability could be written as:

$$\Pr\{r(t+\Delta) = j|r(t) = i\} = \begin{cases} \pi_{ij}\Delta + o(\Delta), & if\ i \neq j, \\ 1 + \pi_{ii}\Delta + o(\Delta), & else, \end{cases} \tag{1.1.4}$$

where π_{ij} represents the transition probability from mode i to mode j, with $\pi_{ij} \geq 0$, $\sum_{j=1}^{l} \pi_{ij} = 1$, and $o(\Delta)$ is the higher order infinitesimal. The matrice $A(r(t))$ and $B(r(t))$ are the functions of the stochastic process $r(t)$, and for each $r(t) = i \in \Xi$, $A(r(t))$ and $B(r(t))$ are real matrices with appropriate dimension.

(2) Mathematical Model of Discrete Generalized Markov Jump Systems

The discrete generalized stochastic Markov jump linear system is described as:

$$Ex(k+1) = A(r(k))x(k) + B(r(k))u(k), \tag{1.1.5}$$

where $E \in \mathbb{R}^{n\times n}$, $x(t) \in \mathbb{R}^n$, $u(t) \in \mathbb{R}^m$ are the same as system (1.1.3), and the elements of the transition probability matrix $\Lambda = [\lambda_{ij}]_{l\times l}$ of the system switching track $r(k) \in \Xi = \{1, \cdots, l\}$ are given by:

$$\lambda_{ij} = \Pr\{r(k+1) = j|r(k) = i\}, \tag{1.1.6}$$

where λ_{ij} represents the transition probability from mode i to mode j, which satisfies that $\lambda_{ij} \geq 0$, $\sum\limits_{j=1}^{l} \lambda_{ij} = 1$.

(3) Applications of Generalized Stochastic Markov Jump Linear Systems

As a special kind of stochastic jump systems, the Markov jump system has practical applications with engineering background. Such as the influence of sudden changes of environment on the behavior of the system, changes of interconnected subsystems, changes of nonlinear system operations, etc., can all be considered as random switching between multimodal systems. Economic system, aircraft control system, robot manipulator system, large space flexible structure system and stochastic decision-making and continuous control systems all have such kinds of system models. Especially in the field of finance and insurance, for example, in 1973, Black and Scholes used geometric Brownian motion to simulate the price of risk assets of options at time t, that is

$$dX(t) = \mu X(t)dt + \sigma X(t)dw(t), \tag{1.1.7}$$

where μ is the rate of return, σ is the disturbance rate, $w(t)$ is the Brownian motion, reflecting the changes of financial market. Although Black used (1.1.7) to give an almost perfect formula of option pricing, the model still had many defects, such as: (a) it failed to depict the discontinuous change of stock price; (b) the empirical analysis showed that the stock volatility was not constant. So many scholars tried to improve the model. On one hand, Merton (1976) put forward a jump diffusion model, which adding a jump process on the model (1.1.7) to characterize the discontinuous changes in stock price [6]. On the other hand, some researchers proposed to let the coefficient of the geometric Brownian motion depends on some hidden Markov chain, that is to say, assuming risk assets are satisfied that:

$$dX(t) = \mu(r(t))X(t)dt + \sigma(r(t))X(t)dw(t), \tag{1.1.8}$$

which $r(t)$ is a Markov chain with finite state, and assuming its state space is $\Xi = \{1, \cdots, l\}$, the infinitesimal operators is $\Pi = [\pi_{ij}]_{l\times l}$. In economics, the state of $r(t)$ is usually called regime-switching or Markov regime-switching, and $X(t)$ is called a process of geometric Brownian motion with Markov regime-switching. The state of $r(t)$ can be interpreted as economic condition structure changes, the regime's replacement, alternating macro news, and economic cycles, etc. There are many literature discussing model (1.1.8), for instance, when $l = 2$, Guo (2000) [7] studied Russia's options pricing problems with Markov modulated geometric Brownian motion model. Guo (2001) [8] further studied an explicit solution to an optimal stopping problem with regime switching, and Jobert (2006) [9] extended the result of Guo into option pricing with finite state Markov-modulated dynamics. Recently, Elliott (2007) [10] studied a class of pricing options under a generalized Markov-modulated jump-diffusion model, assuming that asset price followed:

$$dX(t) = \mu(r(t))X(t)dt + \sigma(r(t))X(t)dw(t) + X(t^-)\int_{\mathbb{R}^+} zN(dt, dz), \tag{1.1.9}$$

where $N(dt, dz)$ is the possion measure.

So, in the field of engineering systems as well as the social and economic systems, such as the option pricing problem in financial engineering, investment insurance dividend distribution problems, multi-sectoral dynamic input-output model of fixed assets, and actual economic system models. All these systems can all be described by the mathematical model of generalized stochastic Markov jump linear systems.

1.1.2 Research Status of Generalized Markov Jump Systems

(1) **Research on the theory of generalized Markov jump systems**

Concrete model of stochastic Markov jump linear systems was first put forward by Krasovskii and Lidskii [11] and Florentin [12] as a numerical example of mathematical analysis. Many researches mainly focused on the stability and stabilization controller design of stochastic jump systems in recent 10 years [13–20]. Professor Mao, one of the famous international scholars in the field of stochastic analysis, issued the asymptotic stability results and numerical methods of stochastic jump systems in his monograph published in 2006 [21]. Professor Mao and his coauthor Dr. Huang studied the stability of singular stochastic Markov jump systems. There are too many researches about the application in engineering and social economy of Markov jump systems, and we can't list them all in a limited space. Our analysis focused on optimal control problem of stochastic Markov jump system (i.e., problem of single stochastic Nash differential game) and the robust control problem which are closely related to this book.

Sworder (1969) first discussed optimal control problem of hybrid linear systems with Markov jump parameters from the perspective of stochastic maximum principle and applied it to the actual control problems [22]. Then, Wonham (1971) proposed the dynamic programming problem of stochastic control system, and successfully applied it to the optimal control of linear jump systems [23]. Fragoso and Costa (2010) gave the separation principle for LQ problems of stochastic Markov jump system in continuous time setting [4]. Görges et al. (2011) proposed the optimal control problems and solution methods of generalized jump systems [5].

Boukas et al. (2001) studied LQR problem of controlled jump rate [24].One of the Chinese scholars named Sun (2006) conducted the control and optimization problem of jump systems, systematically [25]. Mahmoud et al. (2007) gave the analysis results and synthesis of uncertain switched systems in discrete-time setting [26]. Zhang (2009) studied the stability and stabilization of Markov jump linear systems with partly transition probabilities [27–30]. Guo and Gao studied the jump structure control of singular Markov jump systems with time delay [31]. Dong and Gao gave the analysis and control of generalized bilinear Markov jump systems [32]. Zhang and Zhang studied the control theory and application about nonlinear differential algebraic system (including generalized bilinear systems), systematically [33]. Obviously, the research on the singular (or non-singular) stochastic Markov jump linear quadratic optimal control problem (i.e. LQ problem) has relatively obtained a number of achievements, which lay a solid foundation for studying the non-cooperative game theory of generalized stochastic Markov jump systems. But at present, the research results on the LQ non cooperative differential game theory of the generalized stochastic Markov jump system are less, so we put forward the research of LQ non-cooperative differential game theory of generalized stochastic Markov jump systems.

(2) **Research on Non-cooperative differential game theory driven by ordinary differential equations and stochastic differential equations**

The study of game theory has also made abundant achievements, among which, there are many researches on dynamic non-cooperative differential game theory, where the system dynamics are described by differential equations, which includes saddle point equilibrium theory of zero-sum game, Nash equilibrium nonzero-sum game, Stackelberg leader-follower game theory and incentive theory.

For normal systems (such as deterministic and stochastic systems), Basar (1995) [34] summarized the dynamic non-cooperative differential game theory and its application results described by ordinary differential equations and stochastic differential equations in his monograph, systematically (see [34] and cited literatures). Xu and Mizukami have studied the saddle point equilibrium, Nash equilibrium, Stackelberg game theory and incentive theory of linear singular systems, systematically, (see [35–40] and cited literatures). Dockner et al. (2000) described the non-cooperative differential games with its applications, including the capital accumulation, public goods investment, marketing, global pollution control, financial and monetary policy, international trade and other issues of differential games, and this monograph is known as Bible study of differential games [41]. Erickson (2003) introduced the differentia game model of advertising competition, systematically [42]. Zhukovskiy (2003) introduced Lyapunov method in the field of stochastic differential games, in this book, his mainly use the technique of dynamic programming and optimization vector [43]. Jorgensen and Zaccour (2004) mainly studied the differential game in marketing, and introduced the application of differential games in the pricing-making, advertising, marketing channels and other fields [44]. And they had published many research papers with high citations of differential game theory and applications in recent years. Engwerda [45] (2005) introduced the LQ differential game problems and its application examples in economics and management science, and studied the mathematical skills of how to solving the Riccati equations associated with differential games, systematically (Engwerda 2000, 2003; Engwerda and Salmah 2009) [46–48]. Hamadene (1999) studied the nonzero-sum LQ stochastic differential game of BSDEs [49]. The main analytical tools used in these studies are still variation principle, the maximum principle and dynamic planning. In domestic, Academician Zhang Siying's book (1987) [50] "Differential Game" and Professor Li Dengfeng's book(2000) [51] "Differential Game and Its Applications" are the early related literature, but these two books mainly focus on differential games' applications in military and control problems, and pay little attention on applications in economics and management. Because of published influential papers about two zero-sum differential game with impulse control, professor Yong has been highly praised by Berkovitz who is the editor of American Mathematical [52]. Professor Liu also gave the application of leader-follower game to linear multi-sector dynamic input-output of generalized linear system [53]. Wang et al. (2007) studied on the linear quadratic nonzero-sum stochastic differential game under partially observable information [54]; Wu and Yu (2005, 2008) studied the linear quadratic nonzero-sum stochastic differential game problem with stochastic jump, also studied BSDEs differential game with jump and its application in financial engineering [55, 56]; Luo studied the indefinite linear quadratic differential games and indefinite stochastic linear quadratic optimal control problem with Markov jump parameters [57]. In the application of differential games, there is also a growing number of scholars who applying differential game to option pricing [58] (Zheng 2000), the optimal investment in consumption [59, 60] (Liu et al. 1999; Wu and Wu 2001), fisheries resource allocation [61, 62] (Zhang et al. 2000; Zhao et al. 2004), advertising competition and supply chain [63–67] (Zhang and Zhang 2005, 2006; Fu and Zeng 2007, 2008; Xiong etc., 2009), dynamic pricing with network externalities [68] (Liu et al. 2007) and other areas.

(3) **Research on robust control of generalized Markov jump systems**

The results of jump robust control systems are relatively poor. Hespanha (1998) [69] studied the H_∞ control of jump systems. After that, much attention have been paid on H_∞ control. Xu and Chen proposed the H_∞ control of uncertain stochastic bilinear systems with Markov jumps in discrete-time setting [70]. Ting et al. (2010) [71] studied the mixed H_2/H_∞ Robust control problems of stochastic systems with Markov jumps and multiplicative noise in discrete- time setting. All the scholars above used the Lyapunov method (including linear matrix inequality (LMI) method), in this book, we are going to study robust control of stochastic Markov jump linear systems based on game theory. Pioneering work using game theory to study in robust performance controller was first given in the 1960s by Doroto et al. [72], but it did not arouse enough attention due to the need of solving the differential mini-max problem. Since 1990, this design was thought to be used as a powerful weapon to robust design, and the basic idea was transforming the corresponding robust control problem into a two person differential game of saddle point equilibrium or Nash equilibrium. Basar and Limebeer et al. [73, 74] contributed the representative work. And Limebeer et al. converted the mixed H_2/H_∞ control of linear systems into a Nash equilibrium game, and obtained the optimal control strategies. But for the stochastic Markov jump linear systems, there are few results of robust control with various performance based on game theory.

(4) **Research on applications of generalized Markov jump systems**

There are many applications of Markov jump systems in the field of engineering, such as the automatic control of driving shifting systems, traffic management systems and electrical systems, and so on [75]. While application in the field of social science and economic science (in social science and economic science, Markov jump systems are usually referred to Markov switching systems or Markov regime-switching systems) including, ①: risk asset pricing model and the surplus model of an insurer in finance and insurance (detailed description were covered in ref. [76] and reference therein). In terms of VaR measure of risk management in financial market, there exists a fact that the state of financial time series or macroeconomic variables may suddenly change to another state, especially in China's economic entities, based on this fact, doctor Su proposed the ARCH model and the CAPM model with Markov regime-switching, and made the empirical research by using the Chinese data [77, 78]. ②: Dynamic input-output model of multi-sectoral fixed assets (described in detail in the analysis of the socio-economic needs of ①) [79, 80]. ③ Loan pricing of commercial bank with credit rating switching, Dr. Yao represented the credit rating switching process as a continuous time, homogeneous, finite state Markov process, and studied the pricing model of Jarrow et al. (1997) and Lando (1998) [81]. Zhao divided the fluctuation of stock returns in Shanghai Stock Market into three states, "bear", "Bull of mild," "Bull of mad", using the method of MSVAR to exploring the existence of bubbles in stock market, and identified the exact time of speculative bubbles [82].

(5) **Development trend**

Feature 1: At present, more and more special hybrid systems, stochastic Markov jump systems are used to modeling the practical problems in social and economic system, instead of the general stochastic system.

Feature 2: The research on LQ optimal control problem of stochastic Markov jump linear systems has made great process, while the corresponding results of LQ differential game theory are rare.

Trend 1: Analysis and control theory of linear systems can be extended to the analysis and control theory of generalized linear systems; Robust control theory of linear systems can be extended to the generalized linear systems; Dynamic non-cooperative game theory of linear systems can also be extended to the generalized linear systems.

Trend 2: Research on robust control of generalized stochastic Markov jump linear systems has been one of the important research directions, and game theory has become one efficient method to dealing with the robust control problem. Thus, the research on robust control problems of stochastic Markov jump systems based on game theory is a new research direction.

Trend 3: Analysis and control theory of stochastic linear systems can be extended to analysis and control theory of generalized stochastic Markov jump linear systems.

1.2 Differential Games for the Generalized Markov Jump Linear Systems

Differential games study a class of decision problems, under which the evolution of the state is described by a differential equation $\dot{x}(t) = f(t, x(t), u(t), v(t))$, where $u(t)$ and $v(t)$ are control strategies of two players. This differential equation is called the state system of the differential game. When the differential equation is a state equation of generalized systems (also known as singular systems, descriptor systems, or generalized state-space systems: $E\dot{x}(t) = f(t, x(t), u(t), v(t))$, in which E is a known singular matrix with rank(E) = $k \leq n$, and the system is regular. The corresponding differential game is called differential games of the singular systems, and when the differential equation contains random disturbance, that is

$$Edx(t) = f(t, x(t), u(t), v(t))dt + \sigma(t, x(t), u(t), v(t))dw(t), \tag{1.2.1}$$

where $w(t)$ is the random disturbance, the differential game is called the stochastic differential game of generalized dynamic systems. If the behavior of players follows a binding agreement that both players will obey, the game is known as a cooperative game, otherwise known as a non-cooperative game. The book mainly discusses the non-cooperative game, in which each player has an cost function $J_1(t, x(t), u(t), v(t))$ and $J_2(t, x(t), u(t), v(t))$.

When rank(E) = n, that E is nonsingular, (1.2.1) becomes a normal stochastic system, that is

$$dx(t) = \tilde{f}(t, x(t), u(t), v(t)) + \tilde{\sigma}(t, x(t), u(t), v(t))dw(t), \tag{1.2.2}$$

where $\tilde{f}(\cdot,\cdot,\cdot,\cdot) = E \times f(\cdot,\cdot,\cdot,\cdot)$, $\tilde{\sigma}(\cdot,\cdot,\cdot,\cdot) = E \times \sigma(\cdot,\cdot,\cdot,\cdot)$, So it can be said that dynamic non-cooperative differential game theory of a normal (nonsingular) stochastic systems is a special case of that of a generalized (singular) stochastic systems, and the dynamic non-cooperative differential game theory of generalized stochastic systems is the natural generalization of normal stochastic systems.

Non-cooperative differential game theory of the generalized Markov jump linear systems usually contains: the existence conditions and solution methods of equilibrium strategies, such as the saddle-point equilibrium problem, the Nash equilibrium problem and the Stackelberg equilibrium problem. Here, we take the Nash equilibrium problem as an example, the problem is that: how both players choose their strategy control variables $u(t)$ and $v(t)$ to optimizing their cost function $J_1(t, x(t), u(t), v(t))$ and $J_2(t, x(t), u(t), v(t))$, that is to find the control strategy set $(u^*(t), v^*(t))$ and the state $x^*(t)$ satisfies

$$\begin{cases} J_1(t, x^*(t), u^*(t), v^*(t)) \leq J_1(t, x^*(t), u(t), v^*(t)), & \forall u(t) \in \mathcal{U}, \\ J_2(t, x^*(t), u^*(t), v^*(t)) \leq J_2(t, x^*(t), u^*(t), v(t)), & \forall v(t) \in \mathcal{V}, \\ s.t. & \\ E\dot{x}^*(t) = A(r(t))x^*(t) + B(r(t))u^*(t) + C(r(t))v^*(t), & u^*(t) \in \mathcal{U}, v^*(t) \in \mathcal{V}. \end{cases} \tag{1.2.3}$$

Obviously, if there's only one player in the game, the problem of Nash differential games for generalized Markov jump linear systems (1.1.12) becomes an optimal control problem for such systems: to find an optimal control strategy $u^*(t)$ satisfies

$$\begin{cases} \min J(t, x(t), u(t)) \\ s.t. \\ E\dot{x}(t) = A(r(t))x(t) + B(r(t))u(t), \quad u(t) \in \mathcal{U}. \end{cases} \tag{1.2.4}$$

So, the optimal control problem (1.2.4) is a special case of the Nash differential games (1.2.3) for generalized Markov jump linear systems.

1.3 Contents of This Book

Chapter 1. Introduction. This chapter introduces the development and research of the theory for the generalized Markov jump linear system, and the main content of this book.

Chapter 2. The definite and stochastic differential game. This chapter introduces some preliminary knowledge and dynamic optimization technique for the research, and explains basic concepts of the non cooperative differential game and the stochastic differential game.

Chapter 3. The stochastic differential game for the continuous-time Markov jump linear system. This chapter introduces the existence condition, the design and solution of the saddle-point equilibrium strategies, the Nash equilibrium strategies and the Stackelberg strategies for the continuous-time linear Markov jump system.

Chapter 4. The stochastic differential game for the discrete-time Markov jump linear system. This chapter introduces the existence condition, the design and solution of the saddle-point equilibrium strategies, the Nash equilibrium strategies and the Stackelberg strategies for the discrete-time Markov jump linear system.

Chapter 5. The stochastic differential game for the continuous-time generalized Markov linear jump system. This chapter introduces the existence condition, the design method and the approximation algorithm of the saddle-point equilibrium strategies, the Nash equilibrium strategies and the Stackelberg strategies for the continuous-time generalized Markov jump linear system.

Chapter 6. The robust control problems of the generalized Markov jump linear system based on the game theory approach. This chapter studies the H_2/H_∞ robust control problems of the Markov jump linear system. By means of the results of indefinite stochastic differential game for Markov jump linear systems discussed above, we viewed the control strategy designer as one player of the game, i.e. P_1, the stochastic disturbance as another player of the game, i.e. "nature" P_2, respectively, the robust control problems are transformed into a two person differential game model, player P_1 faced the problem that how to design his own strategy in the case of various interference strategy implemented by "nature" P_2, both balanced with the "nature" and optimized his own objective. Corresponding results of stochastic H_∞, H_2/H_∞ control problems for Markov jump linear systems with state, control and disturbance-dependent noise are obtained, and proved the existence of the controller, explicit expressions of the feedback gain are given by means of coupled differential (algebraic) Riccati equations. Finally, numerical examples were presented to verify the validity of the conclusions.

Chapter 7. Applications of stochastic differential game theory for Markov jump linear systems to finance and insurance. A risk minimization problem is considered in a continuous-time Markovian regime switching financial model modulated by a continuous-time, finite-state, Markov chain. We interpret the states of the chain as different states of an economy. A particular form of convex risk measure, which includes the entropic risk measure as a particular case, as a measure of risk and an optimal portfolio is determined by minimizing the convex risk measure of the terminal wealth. We explore the state of the art of the stochastic differential game to formulate the problem as a Markovian regime-switching version of a two-player, zero-sum, stochastic differential game. A novel feature of our model is that we provide the flexibility of controlling both the diffusion risk and the regime-switching risk. A verification theorem for the Hamilton-Jacobi-Bellman (HJB) solution of the game is provided. Furthermore, we studied a game theoretic approach for optimal investment-reinsurance problem of an insurance company under Markovian regime-switching models. In this case, the price dynamics of the underlying risky asset is governed by a Markovian regime switching geometric Brownian motion (GBM). Then, we considered the problem in the context of a two-player, zero-sum stochastic differential game. One of the players in this game is an insurance company and the other is a fictitious player–the market. The insurance company has a utility function and is to select an investment-reinsurance policy, which maximizes the expected utility of the terminal wealth.

References

1. Witsenhausen, H. S. (1966). A class of hybrid-state continuous-time dynamic systems. *IEEE Transactions on Automatic Control, 11*(2), 161–167.
2. Huang, L., & Mao, X. (2011). Stability of singular stochastic systems with Markovian switching. *IEEE Transactions on Automatic Control, 56*(2), 424–429.
3. Deng, F., Luo, Q., & Mao, X. (2012). Stochastic stabilization of hybrid differential equations. *Automatica, 48*(9), 2321–2328.
4. Fragoso, M. D., & Costa, O. L. (2010). A separation principle for the continuous- time LQ-problem with Markovian jump parameters. *IEEE Transactions on Automatic Control, 55* (12), 2692–2707.
5. Görges, D., Izák, M., & Liu, S. (2011). Optimal control and scheduling of switched systems. *Transactions on Automatic Control, 56*(1), 135–140.
6. Merton, R. C. (1976). Option pricing when underlying stock returns are discontinuous. *Journal of Financial Economics, 3*(1), 125–144.
7. Guo, X. (2011). Information and option pricings. *Quantitative Finance, 1*(1), 38–44.
8. Guo, X. (2001). An explicit solution to an optimal stopping problem with regime switching. *Journal of Applied Probability, 38*(2), 464–481.

9. Jobert, A., & Rogers, L. C. (2006). Option pricing with Markov-modulated dynamics. *SIAM Journal on Control and Optimization, 44*(6), 2063–2078.
10. Elliott, R. J., Siu, T. K., Chan, L., & Lau, J. W. (2007). Pricing options under a generalized Markov-modulated jump-diffusion model. *Stochastic Analysis and Applications, 25*(4), 821–843.
11. Krasovskii, N. M., & Lidskii, E. A. (1961). Analytical design of controllers in systems with random attributes. *Automation and Remote Control, 22*(1–3), 1021–1025.
12. Florentin, J. J. (1961). Optimal control of continuous time, Markov, stochastic systems†. *International Journal of Electronics, 10*(6), 473–488.
13. Lin, X.-Z., Z, S.-H., & Yun, Li. (2008). Output feedback stabilization of invariant sets for nonlinear switched systems. *Acta Automatica Sinic, 34*(7), 784–791.
14. Daafouz, J., Riedinger, P., & Iung, C. (2002). Stability analysis and control synthesis for switched systems: A switched Lyapunov function approach. *IEEE Transactions on Automatic Control, 47*(11), 1883–1887.
15. Bacciotti, A. (2002). Stabilization by means of state space depending switching rules. *Systems & Control Letters, 53*(3), 195–201.
16. Cheng, D., Guo, L., Lin, Y., & Wang, Y. (2005). Stabilization of switched linear systems. *IEEE Transactions on Automatic Control, 50*(5), 661–666.
17. Zhao, J., & Dimirovski, G. M. (2004). Quadratic stability of a class of switched nonlinear systems. *IEEE Transactions on Automatic Control, 49*(4), 574–578.
18. Wicks, M., Peleties, P., & DeCarlo, R. (1994, December). Construction of piecewise Lyapunov functions for stabilizing switched systems. In *Proceedings of the 33rd IEEE Conference on Decision and Control, 1994* (Vol. 4, pp. 3492–3497). IEEE.
19. Zhai, G., Lin, H., & Antsaklis, P. J. (2003). Quadratic stabilizability of switched linear systems with polytopic uncertainties. *International Journal of Control, 76*(7), 747–753.
20. Geromel, J. C., Colaneri, P., & Bolzern, P. (2008). Dynamic output feedback control of switched linear systems. *IEEE Transactions on Automatic Control, 53*(3), 720–733.
21. Mao X, Yuan C. (2006). *Stochastic differential equations with Markovian switching*. London: Imperial College Press.
22. Sworder, D. D. (1969). Feedback control of a class of linear systems with jump parameters. *IEEE Transactions on Automatic Control, 14*(1), 9–14.
23. Wonham, W. M. (1970). Random differential equations in control theory. *Probabilistic methods in applied mathematics, 2*(3), 131–212.
24. Boukas, E. K., & Liu, Z. K. (2001). Jump linear quadratic regulator with controlled jump rates. *IEEE Transactions on Automatic Control, 46*(2), 301–305.
25. Sun, Z.-D. (2006). Stabilization and optimization of switched linear systems. *Automatica, 42* (5), 783–788.
26. Mahmoud, M. S., Nounou, M. N., & Nounou, H. N. (2007). Analysis and synthesis of uncertain switched discrete-time systems. *IMA Journal of Mathematical Control and Information*, *24*(2), 245–257.
27. Zhang, L. X., & Boukas, E. K. (2009). Stability and stabilization of Markovian jump linear systems with partly unknown transition probabilities. *Automatica., 45*(2), 463–468.
28. Zhang, L. X., & Boukas, E. K. (2009). Mode-dependent H_∞ filtering for discrete-time Markovian jump linear systems with partly unknown transition probabilities. *Automatica, 45* (6), 1462–1467.
29. Zhang, L. X., & Boukas, E. K. (2009). control for discrete-time Markovian jump linear systems with partly unknown transition probabilities. *International Journal of Robust and Nonlinear Control, 19*(8), 868–883.
30. Zhang, L. X. (2009). Boukas EK. control of a class of extended Markov jump linear systems. *IET Control Theory and Applications, 3*(7), 834–842.

31. Guo, J.-F. (2006). *Variable structure control of time-delay singular systems* (Doctoral dissertation, Ocean University of China).
32. Dong, M. (2008). *Analysis and control for singular bilinear Markovian jump systems* (Doctoral dissertation, Tianjin University).
33. Zhang, X.-H., & Zhang, Q.-L. (2012). Robust passive control for uncertain singular bilinear system. *Journal of Shenyang University (Natural Science), 24*(4), 41–44.
34. Basar, T., & Olsder, G. J. (1999). *Dynamic noncooperative game theory* (2nd edn.). Philadelphia, PA: SIAM.
35. Mizukami, K., & Xu, H. (1992). Closed-loop Stackelberg strategies for linear-quadratic descriptor systems. *Journal of Optimization Theory and Applications, 74*(1), 151–170.
36. Xu, H., & Mizukami, K. (1993). Two-person two-criteria decision making problems for descriptor systems. *Journal of Optimization Theory and Applications, 78*(1), 163–173.
37. Xu, H., & Mizukami, K. (1994). New sufficient conditions for linear feedback closed-loop Stackelberg strategy of descriptor system. *IEEE Transactions on Automatic Control, 39*(5), 1097–1102.
38. Xu, H., & Mizukami, K. (1994). Linear-quadratic zero-sum differential games for generalized state space systems. *IEEE Transactions on Automatic Control, 39*(1), 143–147.
39. Xu, H., & Mizukami, K. (1994). On the Isaacs equation of differential games for descriptor systems. *Journal of Optimization Theory and Applications, 83*(2), 405–419.
40. Xu, H., & Mizukami, K. (1997). Infinite-horizon differential games of singularly perturbed systems: A unified approac. *Automatica, 33*(2), 273–276.
41. Dockner, E., Jørgensen, S., Van Long, N., et al. (2000). *Differential games in economics and management science*. Cambridge: Cambridge University Press.
42. Erickson, G. M. (2003). *Dynamic models of advertising competition*. Boston: Kluwer.
43. Zhukovskiy, V. I. (2003). *Lyapunov functions in differential games*. London: Taylor & Francis.
44. Jørgensen, S., & Zaccour, G. (2004). *Differential games in marketing*. New York: Springer.
45. Engwerda, J. C. (2005). *LQ dynamic optimization and differential games*. New York: Wiley.
46. Engwerda, J. C. (2000). The solution set of the N-player scalar feedback Nash algebraic Riccati equations. *IEEE Transactions on Automatic Control, 45*(12), 2363–2368.
47. Engwerda, J. C. (2003). Solving the scalar feedback Nash algebraic Riccati equations: an eigenvector approach. *IEEE Transactions on Automatic Control, 48*(5), 847–852.
48. Engwerda, J. C., & Salmah, W. I. E. (2009). The open-loop discounted linear quadratic differential game for regular higher order index descriptor systems. *International Journal of Control, 82*(12), 2365–2374.
49. Hamadene, S. (1999). Nonzero sum linear-quadratic stochastic differential games and backward–forward equations. *Stochastic Analysis and Applications, 17*(1), 117–130.
50. Zhang, S.-Y. (1987). *Differential game*. Beijing Science Press.
51. Li, D.-F. (2000). *Differential game and its applications*. Beijing National Defence Industrial Press.
52. Yong, J. (1994). Zero-sum differential games involving impulse controls. *Applied Mathematics and Optimization, 29*(3), 243–261.
53. Liu, X.-P. (1989). *The leader-follower strategies and their extensions in singular systems* (Doctoral dissertation, Northeastern University).
54. Wang, G.-C. (2007). *Partially observed stochastic control systems and their applications* (Doctoral dissertation, Shandong University).
55. Wu, Z., & Yu, Z.-Y. (2005). Linear quadratic nonzero sum differential games with random jumps. *Applied Mathematics and Mechanics*, *8*(26): 945–950.
56. Yu, Z.-Y. (2008). *Some backward problems in stochastic control and game theory* (Doctoral dissertation, Shandong University).
57. Luo, C.-X. (2004). *Indefinite stochastic linear quadratic (LQ) control problems* (Doctoral dissertation, Dalian University of Technology).

58. Zheng, L.-H. (2000). A robust control approach to option pricing. *Journal of Management Science in China, 3*(3), 60–64.
59. Liu, H.-L., & Zheng, L.-H. (1999). The differential game approach for security investment decisions. *Journal of Systems Engineering, 14*(1), 69–72.
60. Liu, H.-L., Wu, C.-F. (2001). Optimal consumption and investment strategy based on worst-case. *Journal of Management Science in China*, 4(6): 48–54.
61. Zhang, D.-X., & Chen, Z.-H. (2000). Examination on differential games to fishery resources quota allocation. *Resources Science, 22*(2), 61–65.
62. Zhao, Z.-Q., Wu, R.-M., & Wang, H.-C. (2004). Optimal operation of fishery resource. *Journal of Systems Engineering, 19*(4), 423–426.
63. Zhang, S.-P., & Zhang, S.-Y. (2005). Stackelberg differential game model for cooperative advertising strategies in supply chain. *Journal of Southwest Jiaotong University, 40*(4), 513–518.
64. Zhang, S.-P., & Zhang, S.-Y. (2006). Dynamic cooperative advertising strategies based on differential games in a supply chain. *Control and Decision, 21*(2), 153–157.
65. Fu, Q., & Zeng, S.-Q. (2007). Differential game models of the vertical cooperative advertising. *Systems Engineering-Theory and Practice, 27*(11), 26–33.
66. Fu, Q., & Zeng, S.-Q. (2008). Game analysis of cooperative advertising and ordering strategies in a supply chain under demand uncertainty. *Systems Engineering-Theory and Practice, 28*(3), 56–63.
67. Xiong, Z.-K., Nie, J.-J., & Li, G.-D. (2009). Competitive brand and generic advertising strategy in oligopoly with differential game. *Journal of Industrial Engineering and Engineering Management, 23*(3), 72–79.
68. Xiong, Z.-K., & Zhang, H.-Y. (2009). Pricing strategy of closed-loop supply chain with asymmetric information. *Industrial Engineering Journal, 12*(003), 39–42.
69. Hespanha, J. P. (1998). *Logic-based switching algorithms in control* (Doctoral dissertation, Yale University).
70. Xu, S., & Chen, T. (2005). Robust H_∞ control for uncertain discrete-time stochastic bilinear systems with Markovian switching. *International Journal of Robust and Nonlinear Control, 15*(5), 201–217.
71. Hou, T., Zhang, W., & Ma, H. (2010). Finite horizon H_2/H_∞ control for discrete-time stochastic systems with Markovian jumps and multiplicative noise. *IEEE Transactions on Automatic Control, 55*(5), 1185–1191.
72. Dorato, P., & Kestenbaum, A. (1967). Application of game theory to the sensitivity design of optimal systems. *IEEE Transactions on Automatic Control, 12*(1), 85–87.
73. Başar, T. (1991). A dynamic games approach to controller design: disturbance rejection in discrete-time. *IEEE Transactions on Automatic Control, 36*(8), 936–952.
74. Limebeer, D. J. N., Anderson, B. D., & Hendel, B. (1994). A Nash game approach to mixed H_2/H_∞ control. *IEEE Transactions on Automatic Control, 39*(1), 69–82.
75. Li, L.-L. (2009). *Study on the H_∞ control problem for classes of switched nonlinear systems* (Doctoral dissertation, Northeastern University).
76. Zhang, Z.-Z. (2009). *The controlled Markov modulated jump diffusion processes and its applications* (Doctoral dissertation, Central South University).
77. Su, T., & Zhan, Y.-R. (2006). VaR estimation based on SWARCH model. *The Journal of Quantitative and Technical Economics, 22*(12), 143–149.
78. Su, T. (2007). *Improved research on VaR measuring method of the financial market risk* (Doctoral dissertation, Tianjin University).
79. Aganovic, Z., Gajic, Z., & Thoma, M. (1995). *Linear optimal control of bilinear systems: With applications to singular perturbations and weak coupling*. Springer, New York, Inc.
80. Zhang, J.-S. (2000). *Computable nonlinear dynamic input output model.* Tsinghua University Press.

81. Yao, X.-H. (2006). *A study on pricing the loan for the commercial bank based on the reduced form models* (Doctoral dissertation, Tianjin University).
82. Zhao, P. (2008). *The analysis on the generation mechanism of speculative bubbles and the empirical study on Chinese stock markets* (Doctoral dissertation, Huazhong University of Science and Technology).

Chapter 2
Deterministic and Stochastic Differential Games

This chapter introduces the theory of deterministic and stochastic differential games, including the dynamic optimization techniques, (stochastic) differential games and their solution concepts, which will lay a foundation for later study.

2.1 Dynamic Optimization Techniques

Consider the dynamic optimization problem in which the single decision-maker:

$$\max_{u}\left\{\int_{t_0}^{T} g[s,x(s),u(s)]ds+q(x(T))\right\}, \tag{2.1}$$

Subject to the vector-valued differential equation:

$$\dot{x}(s)=f[s,x(s),u(s)]ds, \quad x(t_0)=x_0, \tag{2.2}$$

where $x(s)\in X\subset\mathbb{R}^n$ denotes the state variables of game, and $u\in\mathcal{U}$ is the control. The functions $f[s,x,u]$, $g[s,x,u]$ and $q(x)$ are differentiable functions.

Dynamic programming and optimal control are used to identify optimal solutions for the problem (2.1)–(2.2).

2.1.1 *Dynamic Programming*

A frequently adopted approach to dynamic optimization problems is the technique of dynamic programming. The technique was developed by Bellman (1957). The technique is given in Theorem 2.1.1 below.

C.-k. Zhang et al., *Non-cooperative Stochastic Differential Game Theory of Generalized Markov Jump Linear Systems*, Studies in Systems, Decision and Control 67, DOI 10.1007/978-3-319-40587-2_2

Theorem 2.1.1 (Bellman's Dynamic Programming) *A set of controls* $u^*(t) = \phi^*(t,x)$ *constitutes an optimal solution to the control problem (2.1)–(2.2) if there exist continuously differentiable functions* $V(t,s)$ *defined on* $[t_0, T] \times \mathbb{R}^n \to \mathbb{R}$ *and satisfying the following Bellman equation*:

$$\begin{aligned} -V_t(t,x) &= \max_u \{g[t,x,u] + V_x(t,x)f[t,x,u]\} \\ &= \{g[t,x,\phi^*(t,x)] + V_x(t,x)f[t,x,\phi^*(t,x)]\}, \\ V(T,x) &= q(x). \end{aligned}$$

Proof Define the maximized payoff at time t with current state x as a value function in the form:

$$\begin{aligned} V(t,x) &= \max_u \left[\int_t^T g(s,x(s),u(s))ds + q(x(T))\right] \\ &= \int_t^T g[s,x^*(s),\phi^*(s,x^*(s))]ds + q(x^*(T)). \end{aligned}$$

Satisfying the boundary condition

$$V(T,x^*(T)) = q(x^*(T)),$$

and

$$\dot{x}^*(s) = f[s,x^*(s),\phi^*(s,x^*(s))], \quad x^*(t_0) = x_0.$$

If in addition to $u^*(s) \equiv \phi^*(s,x)$, we are given another set of strategies, $u(s) \in \mathcal{U}$, with the corresponding terminating trajectory $x(s)$, then Theorem 2.1.1 implies

$$\begin{aligned} g(t,x,u) + V_x(t,x)f(t,x,u) + V_t(t,x) &\le 0, \\ g(t,x^*,u^*) + V_{x^*}(t,x^*)f(t,x^*,u^*) + V_t(t,x^*) &= 0. \end{aligned}$$

Integrating the above expressions from t_0 to T, we obtain

$$\begin{aligned} &\int_{t_0}^T g(s,x(s),u(s))ds + V(T,x(T)) - V(t_0,x_0) \le 0, \\ &\int_{t_0}^T g(s,x^*(s),u^*(s))ds + V(T,x^*(T)) - V(t_0,x_0) \le 0. \end{aligned}$$

Elimination of $V(t_0, x_0)$ yields

$$\int_{t_0}^{T} g(s, x(s), u(s))ds + q(x(T)) \leq \int_{t_0}^{T} g(s, x^*(s), u^*(s))ds + q(x^*(T)).$$

From which it readily follows that u^* is the optimal strategy.

Upon substituting the optimal strategy $\phi^*(t, x)$ into (2.2) yields the dynamics of optimal state trajectory as:

$$\dot{x}(s) = f[s, x(s), \phi^*(s, x(s))]ds, \quad x(t_0) = x_0. \tag{2.3}$$

Let $x^*(t)$ denote the solution to (2.3). The optimal trajectory $\{x^*(t)\}_{t=t_0}^{T}$ can be expressed as:

$$x^*(t) = x_0 + \int_{t_0}^{t} f[s, x^*(s), \phi^*(s, x^*(s))]ds. \tag{2.4}$$

For notational convenience, we use the terms $x^*(t)$ and x_t^* interchangeably. The value function $V(t, x)$ where $x = x_t^*$ can be expressed as

$$V(t, x) = \int_{t}^{T} g[s, x^*(s), \phi^*(s)]ds + q(x^*(T)).$$

2.1.2 Optimal Control

The maximum principle of optimal control was developed by Pontryagin (details in Pontryagin et al (1962)). Consider again the dynamic optimization problem (2.1)–(2.2).

Theorem 2.1.2 (Pontryagin's Maximum Principle) *A set of controls $u^*(s) = \zeta^*(s, x_0)$ provides an optimal solution to control problem (2.1)–(2.2), and $\{x^*(s), t_0 \leq s \leq T\}$ is the corresponding state trajectory, if there exist costate functions $\Lambda(s) : [t_0, T] \to \mathbb{R}^m$ such that the following relations are satisfied:*

$$\begin{aligned}
\zeta^*(s, x_0) \equiv u^*(s) &= \arg\max\{g[s, x^*(s), u(s)] + \Lambda(s) f[s, x^*(s), u(s)]\}, \\
\dot{x}^*(s) &= f[s, x^*(s), u^*(s)], \quad x^*(t_0) = x_0, \\
\dot{\Lambda}(s) &= -\frac{\partial}{\partial x}\{g[s, x^*(s), u^*(s)] + \Lambda(s) f[s, x^*(s), u^*(s)]\}, \\
\Lambda(T) &= \frac{\partial}{\partial x^*} q(x^*(T)).
\end{aligned}$$

Proof First define the function (Hamiltonian)

$$H(t,x,u) = g(t,s,u) + V_x(t,x)f(t,x,u).$$

From Theorem 2.1.2, we obtain

$$-V_t(t,x) = \max_u H(t,x,u).$$

This yields the first condition of Theorem 2.1.2. Using u^* to denote the payoff maximizing control, we obtain

$$H(t,x,u^*) + V_t(t,x){=}0.$$

Which is an identity in x. Differentiating this identity partially with respect to x yields

$$V_{tx}(t,x) + g_x(t,x,u^*) + V_x(t,x)f_x(t,x,u^*) + V_{xx}(t,x)f(t,x,u^*)$$
$$+ [g_u(t,s,u) + V_x(t,x)f_u(t,x,u^*)]\frac{\partial u^*}{\partial x} = 0.$$

If u^* is an interior point, then $[g_u(t,x,u^*) + V_x(t,x)f_u(t,x,u^*)] = 0$ according to the condition $-V_t(t,x) = \max_u H(t,x,u)$. If u^* is not an interior point, then it can be shown that

$$[g_u(t,x,u^*) + V_x(t,x)f_u(t,x,u^*)]\frac{\partial u^*}{\partial x} = 0.$$

(because of optimality, $[g_u(t,x,u^*) + V_x(t,x)f_u(t,x,u^*)]$ and $\frac{\partial u^*}{\partial x}$ are orthogonal; and for specific problems we may have $\frac{\partial u^*}{\partial x} = 0$). Moreover, the expression $V_{tx}(t,x) + V_{xx}(t,x)f(t,x,u^*) \equiv V_{tx}(t,x) + V_{xx}(t,x)\dot{x}$ can be written as $[dV_x(t,x)](dt)^{-1}$. Hence, we obtain:

$$\frac{dV_x(t,x)}{dt} + g_x(t,x,u^*) + V_x(t,x)f_u(t,x,u^*) = 0.$$

By introducing the costate vector, $\Lambda(t) = V_{x^*}(t,x^*)$, where $x*$ denotes the state trajectory corresponding to u^*, we arrive at

$$\frac{dV_x(t,x^*)}{dt} = \dot{\Lambda}(s) = -\frac{\partial}{\partial x}\{g[s,x^*(s),u^*(s)] + \Lambda(s)f[s,x^*(s),u^*(s)]\}.$$

Finally, the boundary condition for $\Lambda(t)$ is determined from the terminal condition of optimal control in Theorem 2.1.2 as

$$\Lambda(T) = \frac{\partial V(T, x^*)}{\partial x} = \frac{\partial q(x^*)}{\partial x}.$$

Then,we obtain Theorem 2.1.2.

2.1.3 Stochastic Control

Consider the dynamic optimization problem in which the single decision maker

$$\max_u \mathbf{E}_{t_0}\left\{\int_{t_0}^{T} g[s, x(s), u(s)]ds + q(x(T))\right\}, \tag{2.5}$$

Subject to the vector-valued stochastic differential equation:

$$dx(s) = f[s, x(s), u(s)]ds + \sigma[s, x(s)]dw(s), \quad x(t_0) = x_0, \tag{2.6}$$

where $\mathbf{E}_{t_0}$ denotes the expectation operator performed at time t_0, and $\sigma[s, x(s)]$ is a $n \times \Theta$ matrix and $w(s)$ is a Θ dimensional Brownian motion and the initial state x_0 is given. Let $\Omega[s, x(s)] = \sigma[s, x(s)]\sigma[s, x(s)]'$ denote the covariance matrix with its element in row h and column ζ denoted by $\Omega^{h\zeta}[s, x(s)]$.

The technique of stochastic control developed by Fleming (1969) can be applied to solve the problem.

Theorem 2.1.5 *A set of controls $u^*(t) = \phi^*(t, x)$ constitutes an optimal solution to the problem (2.5)–(2.6), if there exist continuously differentiable functions $V(t, s)$ $[t_0, T] \times \mathbb{R}^n \to \mathbb{R}$, satisfying the following partial differential equation:*

$$- V_t(t, x) - \frac{1}{2}\sum_{h,\zeta=1}^{n} \Omega^{h\zeta}(t, x)V_{x^h x^\zeta}(t, s) = \max_u\{g\phi[t, x,u] + V_x f[t, x, u]\},$$
$$V(T, x) = q(x).$$

Proof Substitute the optimal control $\phi^*(t, x)$ into the (2.6) to obtain the optimal state dynamics as

$$\begin{aligned} &dx(s) = f[s, x(s), \phi^*(s, x(s))]ds + \sigma[s, x(s)]dw(s), \\ &x(t_0) = x_0. \end{aligned} \tag{2.7}$$

The solution to (2.7), denoted by $x^*(t)$, can be expressed as

$$x^*(t) = x_0 + \int_{t_0}^{t} f[s, x^*(t), \phi^*(s, x^*(t))]ds + \int_{t_0}^{T} \sigma[s, x^*(t)]dw(s). \tag{2.8}$$

We use X_t^* to denote the set of realizable values of x_t^* at time t generated by (2.8). The term x_t^* is used to denote an element in the set x_t^*.

Define the maximized payoff at time t with current state x_t^* as a value function in the form

$$V(t, x_t^*) = \max_u \mathbf{E}_{t_0}\left\{\int_t^T g(s, x(s), u(s))ds + q(x(T))|x(t) = x_t^*\right\}$$
$$= \mathbf{E}_{t_0}\int_t^T g[s, x^*(s), \phi^*(s, x^*(s))]ds + q(x^*(T)).$$

Satisfying the boundary condition

$$V(T, x^*(T)) = q(x^*(T)).$$

One can express $V(t, x_t^*)$ as

$$V(t, x_t^*) = \max_u \mathbf{E}_{t_0}\left\{\int_t^T g(s, x(s), u(s))ds + q(x(T))|x(t) = x_t^*\right\}$$
$$= \max_u \mathbf{E}_{t_0}\left\{\int_t^{t+\Delta t} g(s, x(s), u(s))ds + V(t + \Delta t, x_t^* + \Delta x_t^*)|x(t) = x_t^*\right\}. \tag{2.9}$$

where

$$\Delta x_t^* = f[t, x_t^*, \phi^*(t, x_t^*)]\Delta t + \sigma[t, x_t^*]\Delta z_t + o(\Delta t),$$
$$\Delta w_t = w(t + \Delta t) - w(t).$$

With $\Delta t \to 0$, applying Ito's lemma Eq. (2.9) can be expressed as:

$$V(t, x_t^*) = \max_u E_{t_0}\{g[t, x_t^*, u]\Delta t + V(t, x_t^*) + V(t, x_t^*)\Delta t + V_{x_t}(t, x_t^*)f[t, x_t^*, \phi^*(t, x_t^*)]\Delta t + V_{x_t}(t, x_t^*)\sigma[t, x_t^*]\Delta w + \frac{1}{2}\sum_{h,\zeta=1}^{m}\Omega^{h\zeta}(t, x)V_{x^h x^\zeta}(t, x)\Delta t + o(\Delta t)\}. \tag{2.10}$$

Dividing (2.10) throughout by Δt, with $\Delta t \to 0$, and taking expectation yields

$$- V(t, x_t^*) - \frac{1}{2}\sum_{h,\zeta=1}^{m} \Omega^{h\zeta}(t,x) V_{x^h x^\zeta}(t,x) =$$
$$= \max_u \{ g[t, x_t^*, u] + V_{x_t}(t, x_t^*) f[t, x_t^*, \phi^*(t, x_t^*)] \Delta t + V_{x_t}(t, x_t^*) \}.$$

With boundary condition

$$V(T, x^*(T)) = q(x^*(T)).$$

2.2 Differential Games and Their Solution Concepts

Firstly we introduce the definition of differential game briefly:

Definition 2.2.1 If the time difference between each phase of the game narrowed to the minimum limit, differential games can be considered as continuous-time dynamic games. A continuous-time infinite dynamic games of the initial state x_0 and continuous time $T - t_0$, and can be expressed as $\Gamma(x_0, T - t_0)$.

In particular, in the general n-person differential game, Player i seek to:

$$\max_{u_i} \int_{t_0}^{T} g^i[s, x(s), u_1(s), \cdots, u_n(s)] ds + q^i(x(T)). \tag{2.11}$$

For $i \in N = \{1, 2, \cdots, n\}$, where $g^i(\cdot) \geq 0$ and $q^i(\cdot) \geq 0$.
Subject to the deterministic dynamics

$$\dot{x}(s) = f[s, x(s), u_1(s), \ldots, u_n(s)], x(t_0) = x_0. \tag{2.12}$$

The functions $f[s, x(s), u_1(s), \ldots, u_n(s)]$, $g^i[s, x(s), u_1(s), u_2(s), \cdots, u_n(s)]$ and $q^i(\cdot)$, for $i \in N$, $s \in [t_0, T]$ are differentiable functions.

2.2.1 *Open-Loop Nash Equilibria*

If the players choose to commit their strategies from the outset, the players' information structure can be seen as an open-loop pattern in which $\eta^i(s) = \{x_0\}, s \in [t_0, T]$. Their strategies become functions of the initial state x_0 and time s, and can be expressed as $\{\mu_i(s) = \vartheta_i(s, x_0)\}$, for $i \in N$. An open-loop Nash equilibrium for the game is characterized as follows.

Theorem 2.2.1 *For the differential game (2.11) and (2.12), a set of strategies* $\{u_i^*(s) = \zeta_i^*(s, x_0), i \in N\}$ *provides an open-loop Nash equilibrium, an* $\{x^*(s),\ t_0 \le s \le T\}$ *is the corresponding state trajectory, if there exist n costate functions* $\Lambda^i(s) : [t_0, T] \to \mathbb{R}^n$, *for* $i \in N$, *such that the following relations are satisfied*:

$$\zeta_i^*(s, x_0) \equiv u_i^*(s) = \arg\max_{u_i \in \mathcal{U}^i} \{g^i[s, x^*(s), u_1^*(s), \cdots, u_{i-1}^*(s), u_i(s), u_{i+1}^*(s) \cdots, u_n^*(s)]$$

$$+ \Lambda^i(s) f[s, x^*(s), u_1^*(s), \cdots, u_{i-1}^*(s), u_i(s), u_{i+1}^*(s) \cdots, u_n^*(s)]\},$$

$$\dot{x}^*(s) = f[s, x^*(s), u_1^*(s), \cdots, u_n^*(s)], x^*(t_0) = x_0,$$

$$\dot{\Lambda}^i(s) = -\frac{\partial}{\partial x^*} \{g^i[s, x^*(s), u_1^*(s), \cdots, u_n^*(s)] + \Lambda^i(s) f[s, x^*(s), u_1^*(s), \cdots, u_n^*(s)]\}.$$

According to the analysis above, we know that:

First, given the optimal strategies of players, they should maximize the sum of the instantaneous payment and integration of state variation and covariate function in current time at every time point. That is, not only the instantaneous payment but also the whole payment influenced by state variation should be considered when one player chooses the optimal strategy. Second, the variation of optimal state depends on the optimal strategies of all the players, current time and state, and the optimal state of the beginning consistent with the initial state of the game. Third, given the optimal strategies of players $i \in N$ which only depend on current time and initial state, the variation of covariate functions depend on current instantaneous payment, variation of current state and current covariate functions. The value of covariate function equal to the marginal impact of optimal state at the end of game. Therefore, covariate functions of players reflect the impacts on future payment by the variation of optimal state.

2.2.2 Closed-Loop Nash Equilibria

After discussing the necessary conditions of open-loop Nash Equilibria, then we study the necessary conditions of closed-loop Nash Equilibria.

The players' information structures follow the pattern $\eta^i(s) = \{x_0, x(s)\}, s \in [t_0, T]$, for $i \in N$. The players' strategies become functions of the initial state x_0, current state $x(s)$ and current time s, and can be expressed as $\{u_i(s) = \vartheta_i(s, x(s), x_0), i \in N\}$. The following theorem provides a set of necessary conditions for any closed-loop no-memory Nash equilibrium solution to satisfy.

Theorem 2.2.2 *A set of strategies* $\{u_i(s) = \vartheta_i(s, x, x_0), i \in N\}$ *provides a closed-loop no memory Nash equilibrium solution to the game* (2.11)−(2.12), *and* $\{x^*(s), t_0 \le s \le T\}$ *is the corresponding state trajectory, if there exist n costate functions* $\Lambda^i(s) : [t_0, T] \to \mathbb{R}^n$, *for* $i \in N$, *such that the following relations are satisfied*:

$$\vartheta_i^*(s,x^*,x_0) \equiv u_i^*(s) = \arg\max_{u_i \in \mathcal{U}^i}\{g^i[s,x^*(s),u_1^*(s),\cdots,u_{i-1}^*(s),u_i(s),u_{i+1}^*(s)\cdots,u_n^*(s)]$$
$$+\Lambda^i(s)f[s,x^*(s),u_1^*(s),\cdots,u_{i-1}^*(s),u_i(s),u_{i+1}^*(s)\cdots,u_n^*(s)]\},$$
$$\dot{x}^*(s) = f[s,x^*(s),u_1^*(s),\cdots,u_n^*(s)],x^*(t_0) = x_0,$$
$$\dot{\Lambda}^i(s) =$$
$$-\frac{\partial}{\partial x^*}\{g^i[s,x^*(s),\vartheta_1^*(s,x^*,x_0),\cdots,\vartheta_{i-1}^*(s,x^*,x_0),$$
$$u_i^*(s),\vartheta_{i+1}^*(s,x^*,x_0),\cdots\vartheta_n^*(s,x^*,x_0)] + \Lambda^i(s)f[s,x^*(s),\vartheta_1^*(s,x^*,x_0),\cdots,$$
$$\vartheta_{i-1}^*(s,x^*,x_0),u_i^*(s),\vartheta_{i+1}^*(s,x^*,x_0),\cdots\vartheta_n^*(s,x^*,x_0)]\},$$
$$\Lambda^i(T) = \frac{\partial}{\partial x^*}q^i(x^*(T)), i \in N.$$

Then a set of strategies $\{u_i(s) = \vartheta_i(s,x,x_0), i \in N\}$ *provides a closed-loop no memory Nash equilibrium.*

According to Theorem 2.2.2, similar to the open-loop situation, in closed-loop Nash equilibrium solution, we know that:

First, given the optimal strategies of players, they should maximize the sum of the instantaneous payment and integration of state variation and covariate function in current time at every time point. That is, not only the instantaneous payment but also the whole payment influenced by state variation should be considered when one player chooses the optimal strategy. Second, the variation of optimal state depends on the optimal strategies of all the players, current time and state, and the optimal state of the beginning consistent with the initial state of the game. Third, given the optimal strategies of players $i \in N$ which only depend on current time and initial state, the variation of covariate functions depend on current instantaneous payment, variation of current state and current covariate functions. The value of covariate function equal to the marginal impact of optimal state at the end of game. Therefore, covariate functions of players reflect the impacts on future payment by the variation of optimal state. Note that the partial derivatives of covariate function on optimal state depend on strategies of other players.

2.2.3 Feedback Nash Equilibria

The set of equations of closed-loop Nash Equilibria in general admits of an uncountable number of solutions, which correspond to "informationally non-unique" Nash equilibrium solutions of differential games under memoryless perfect state information pattern. Derivation of nonunique closed-loop Nash equilibria can be found in Mehlmann and Willing (1984). To eliminate information nonunique-ness in the derivation of Nash equilibria, one can constrain the Nash solution further by requiring it to satisfy the feedback Nash equilibrium property. In particular, the players' information structures follow either a closed-loop perfect state (CLPS) pattern in which $\eta^i(s) = \{x(t), t_0 \leq t \leq s\}$ or amemoryless perfect state (MPS) pattern in which $\eta^i(s) = \{x_0, x(s)\}$. Moreover, we require the following feedback Nash equilibrium condition to be satisfied.

Definition 2.3 For the n-person differential game (2.11)–(2.12), with MPS or CLPS information, an n-tuple of strategies $\{u_i^*(s) = \phi_i^*(s,x) \in \mathcal{U}^i, i \in N\}$ constitutes a feedback Nash equilibrium solution if there exist functionals $V^i(t,x), i \in N$ defined on $[t_0, T] \times \mathbb{R}^n$ and satisfying the following relations:

$$V^i(t,x) = \int_t^T g^i[s, x^*(s), \phi_1^*(s,\eta_s), \cdots, \phi_n^*(s,\eta_s)]ds + q^i(x^*(T)) \geq$$

$$\int_t^T g^i[s, x^{[i]}(s), \phi_1^*(s,\eta_s), \cdots, \phi_{i-1}^*(s,\eta_s), \phi_i(s,\eta_s), \phi_{i+1}^*(s,\eta_s), \cdots, \phi_n^*(s,\eta_s)]ds$$

$$+ q^i(x^{[i]}(T)), \forall \phi_i(\cdot,\cdot) \in \mathcal{U}^i, x \in \mathbb{R}^n,$$

$$V^i(T,x) = q^i(x)$$

where on the interval $[t_0, T]$,

$$x^{[i]}(t) = x,$$

$$\dot{x}^*(s) = f[s, x^*(s), \phi_1^*(s,\eta_s), \cdots, \phi_n^*(s,\eta_s)], x(s) = x,$$

$\eta(s)$ stands for either the data set $\{x(s), x_0\}$ or $\{x(\tau), \tau \leq s\}$, depending on whether the information pattern is MPS or CLPS. Therefore the players' strategies can be expressed as $\{u_i^*(s) = \phi_i^*(s,x) \in U^i, i \in N\}$.

The following theorem provides a set of necessary conditions characterizing a feedback Nash equilibrium solution for the game (2.11)–(2.12) is characterized as follows:

Theorem 2.2.3 *An n-tuple of strategies* $\{u_i^*(t) = \phi_i^*(t,x) \in \mathcal{U}^i, i \in N\}$ *provides a feedback Nash equilibrium solution to the game* (2.11)–(2.12) *if there exist continuously differentiable functions* $V^i(t,x) : [t_0, T] \times \mathbb{R}^n \to \mathbb{R}$, *i*∈*N, satisfying the following set of partial differential equations*:

$$\begin{aligned} -V_t^i(t,x) &= \max_{u_i}\{g^i[t, x, \phi_1^*(t,x), \ldots, \phi_{i-1}^*(t,x), u_i(t,x), \phi_{i+1}^*(t,x)\ldots, \varphi_n^*(t,x)] \\ &\quad + V_x^i(t,x) f[t, x, \phi_1^*(t,x), \ldots, \phi_{i-1}^*(t,x), u_i(t,x), \phi_{i+1}^*(t,x), \ldots, \phi_n^*(t,x)]\} \\ &= \{g^i[t, x, \phi_1^*(t,x), \ldots, \varphi_n^*(t,x)] + V_x^i(t,x) f[t, x, \phi_1^*(t,x), \ldots, \phi_n^*(t,x)]\}, \\ V^i(T,x) &= q^i(x), i \in N. \end{aligned}$$

Theorem 2.2.4 *A pair of strategies* $\{\phi_i^*(t,x); i = 1,2\}$ *provides a feedback saddle-point solution to the zero-sum version of the game (2.11)–(2.12) if there exists a function* $V : [t_0, T] \times \mathbb{R}^n \to \mathbb{R}$ *satisfying the partial differential equation*:

$$\begin{aligned}-V_t(t,x) &= \min_{u_1\in\mathcal{U}^1}\max_{u_2\in\mathcal{U}^2}\{g[t,x,u_1(t),u_2(t)]+V_xf[t,x,u_1(t),u_2(t)]\}\\ &= \max_{u_2\in\mathcal{U}^2}\min_{u_1\in\mathcal{U}^1}\{g[t,x,u_1(t),u_2(t)]+V_xf[t,x,u_1(t),u_2(t)]\}\\ &= \{g[t,x,\phi_1^*(t,x),\phi_2^*(t,x)]+V_xf[t,x,\phi_1^*(t,x),\phi_2^*(t,x)]\},\\ V(T,x) &= q(x).\end{aligned}$$

According to the necessary condition of feedback Nash equilibrium solution, there are two points should to note,

First, the value of the value functions of each player will change as time when they choose the optimal strategies under current time and state. Second, the payments of each player at the last time point are equal to that in the end of game.

2.3 Stochastic Differential Games and Their Solutions

We introduce the deterministic differential games and their solutions with stochastic factors.

2.3.1 *The Model of Stochastic Differential Game*

One way to incorporate stochastic elements in differential games is to introduce stochastic dynamics. A stochastic formulation for quantitative differential games of prescribed duration involves a vector-valued stochastic differential equation

$$\begin{aligned}&dx(s)=f[s,x(s),u_1(s),u_2(s),\ldots,u_n(s)]ds+\sigma[s,x(s)]dw(s),\\ &x(t_0)=x_0.\end{aligned} \tag{2.13}$$

which describes the evolution of the state and N objective functionals

$$\mathbf{E}_{t_0}\left\{\int_{t_0}^{T} g^i[s,x(s),u_1(s),u_2(s),\ldots,u_n(s)]ds+q^i(x(T))\right\}, i\in N \tag{2.14}$$

with $\mathbf{E}_{t_0}\{\cdot\}$ denoting the expectation operation taken at time t_0, $\sigma[s,x(s)]$ is a $n\times\Theta$ matrix and $w(s)$ is a Θ dimensional Brownian motion and the initial state x_0 is given. Let $\Omega[s,x(s)]=\sigma[s,x(s)]\sigma[s,x(s)]'$ denote the covariance matrix with its element in row h and column ζ denoted by $\Omega^{h\zeta}[s,x(s)]$. Moreover, $\mathbf{E}[dw_\varpi]=0$, $\mathbf{E}[dw_\varpi dt]=0$, and $\mathbf{E}\left[(dw_\varpi)^2\right]=dt$, for $\varpi\in[1,2,\ldots,\Theta]$; $\mathbf{E}[dw_\varpi dw_\omega]=0$, for $\varpi\in[1,2,\ldots,\Theta]$, $\omega\in[1,2,\ldots,\Theta]$ and $\varpi\neq\omega$. Given the stochastic nature, the information structures must follow the MPS pattern or CLPS pattern or the feedback perfect state (FB) pattern in which $\eta^i(s)=\{x(s)\}$, $s\in[t_0,T]$.

2.3.2 The Solutions of Stochastic Differential Game

The character of stochastic differential game is the state changes with the stochastic dynamic system in every moment. Therefore, stochastic differential game is closer to reality compared with the deterministic differential game. Based on this, the following section only discuss the feedback solutions which are more realistic then the open-loop solution. A Nash equilibrium of the stochastic game (2.13)−(2.14) can be characterized as:

Theorem 2.3.1 *An n-tuple of feedback strategies* $\left\{\phi_i^*(t,x)\in\mathcal{U}^i; i\in N\right\}$ *provides a Nash equilibrium solution to the game (2.13)−(2.14) if there exist suitably smooth functions* $V^i:[t_0,T]\times\mathbb{R}^n\to\mathbb{R}$, *satisfying the semilinear parabolic partial differential equations*

$$\begin{aligned}
-V_t^i-\frac{1}{2}\sum_{h,\zeta}^{n}\Omega^{h\zeta}(t,x)V^i_{x_hx_\zeta} &= \max_{u_i}\left\{g^i\left[t,x,\phi_1^*(t,x),\ldots,\phi_{i-1}^*(t,x),u_i(t),\phi_{i+1}^*(t,x),\ldots,\phi_n^*(t,x)\right]\right.\\
&\quad\left.+V_x^i(t,x)f\left[t,x,\phi_1^*(t,x),\ldots,\phi_{i-1}^*(t,x),u_i(t),\phi_{i+1}^*(t,x),\ldots,\phi_n^*(t,x)\right]\right\}\\
&=\left\{g^i\left[t,x,\phi_1^*(t,x),\phi_2^*(t,x),\ldots,\phi_n^*(t,x)+V_x^i(t,x)f\left[t,x,\phi_1^*(t,x),\ldots,\phi_n^*(t,x)\right]\right\},\\
V^i(T,x) &= q^i(x), i\in N.
\end{aligned}$$

Proof This result follows readily from the definition of Nash equilibrium and from Theorem 2.1.2, since by fixing all players' strategies, except the ith one's, at their equilibrium choices (which are known to be feedback by hypothesis), we arrive at a stochastic optimal control problem of the type covered by Theorem 2.3.1 and whose optimal solution (if it exists) is a feedback strategy.

Consider the two-person zero-sum version of the game (2.13)–(2.14) in which the payoff of Player 1 is the negative of that of Player 2. Under either MPS or CLPS information pattern, a Nash equilibrium solution can be characterized as follows.

Theorem 2.3.2 *A pair of strategies* $\left\{\phi_i^*(t,x)\in\mathcal{U}^i; i=1,2\right\}$ *provides a feedback saddle-point solution to the two-person zero-sum version of the game (2.13)–(2.14) if there exists a function* $\Lambda(s):[t_0,T]\to\mathbb{R}^n$ *satisfying the partial differential equation*:

$$\begin{aligned}
-V_t-\frac{1}{2}\sum_{h,\zeta}^{n}\Omega^{h\zeta}(t,x)V_{x_hx_\zeta} &= \min_{u_1\in\mathcal{U}^1}\max_{u_2\in\mathcal{U}^2}\left\{g[t,x,u_1,u_2]+V_xf[t,x,u_1,u_2]\right\}\\
&=\max_{u_2\in\mathcal{U}^2}\min_{u_1\in\mathcal{U}^1}\left\{g[t,x,u_1,u_2]+V_xf[t,x,u_1,u_2]\right\}\\
&=\left\{g\left[t,x,\phi_1^*(t,x),\phi_2^*(t,x)\right]+V_xf\left[t,x,\phi_1^*(t,x),\phi_2^*(t,x)\right]\right\},\\
V(T,x) &= q(x).
\end{aligned}$$

Proof This result follows as a special case of Theorem 2.3.1 by taking $n=2$, $g^1(\cdot)=-g^2(\cdot)\equiv g(\cdot)$, and $q^1(\cdot)=-q^2(\cdot)\equiv q(\cdot)$, in which case $V^1=-V^2\equiv V$

and existence of a saddle point is equivalent to interchangeability of the min max operations.

According to the necessary condition of feedback Nash Equilibria, there are two points we should to know,

First, the value functions in stochastic differential game (2.13)–(2.14) change with time when all the players (include i) determine the optimal strategies depend on current time and state. Second, the value function of player $i \in N$ in last point equals to his final payment in the game.

Chapter 3
Stochastic Differential Games of Continuous-Time Markov Jump Linear Systems

This chapter mainly discussed the stochastic differential game theory of continuous-time Markov jump linear systems. Firstly, the stochastic LQ problem of Markov jump linear systems was reviewed. Then, two person nonzero Nash games in finite-time horizon and infinite-time horizon were discussed, and the existence conditions and strategy design method of equilibrium strategies were given. Finally, the Stackelberg game problem with two players was studied, and the existence conditions for the Stackelberg equilibrium strategy were obtained.

3.1 Stochastic LQ Problem—Differential Game with One Player

The LQ control with Markovian jumps has been very widely studied for the last two decade:

$$\begin{aligned}
&\text{minimize} \quad J = \mathbf{E}\left\{\int_0^T \left[x(t)'Q(t,r_t)x(t) + u(t)'R(t,r_t)u(t)\right]dt \right.\\
&\qquad\qquad\qquad \left. + x(T)'Hx(T)|r_0 = i\right\}, \\
&s.t.\\
&\begin{cases} dx(t) = [A(t,r_t)x(t) + B(t,r_t)u(t)]dt + \sigma(t,r_t)dw(t),\\ x(0) = x_0 \in \mathbb{R}^n,\end{cases}
\end{aligned} \tag{3.1.1}$$

where r_t is a Markov chain taking values in $\{1,\ldots,l\}$, $w(t)$ is a standard Brownian motion independent of r_t, and $A(t,r_t) = A_i(t)$, $B(t,r_t) = B_i(t)$, $\sigma(t,r_t) = \sigma_i(t)$, $Q(t,r_t) = Q_i(t)$ and $R(t,r_t) = R_i(t)$ when $r_t = i$ $(i = 1,\ldots,l)$. Here the matrix functions $A_i(\cdot)$, etc. are given with appropriate dimensions. The Markov chain r_t has the transition probabilities given by:

C.-k. Zhang et al., *Non-cooperative Stochastic Differential Game Theory of Generalized Markov Jump Linear Systems*, Studies in Systems, Decision and Control 67, DOI 10.1007/978-3-319-40587-2_3

$$\Pr\{r_{t+\Delta}=j|r_t=i\}=\begin{cases}\pi_{ij}\Delta+o(\Delta), & \textit{if } i\neq j,\\ 1+\pi_{ij}\Delta+o(\Delta), & \textit{else},\end{cases} \tag{3.1.2}$$

where $\pi_{ij}\geq 0$ for $i\neq j$ and $\pi_{ii}=-\sum_{i\neq j}\pi_{ij}$.

In (3.1.1), when the diffusion term $\sigma(t,r_t)$ does not include $u(t)$, it is usually required that the state weighting matrices, $Q_i(t)$ and the control weighting matrices, $R_i(t)$ be positive semidefinite and positive definite, respectively. But when $u(t)$ is included in the diffusion term $\sigma(t,r_t)$, the control weighting matrices, $R_i(t)$ in the cost function J can be indefinite, and academics call it the indefinite stochastic LQ problem that has been widely used in reality, especially in the field of mathematical finance.

For completeness content, and laying the foundation for the later study, this section discusses the stochastic LQ control of continuous time Markov jump linear system with state and control included in the diffusion term.

3.1.1 Finite-Time Horizon Case

3.1.1.1 Problem Formulation

Given a filtered probability space $(\Omega,\mathcal{F},\{\mathcal{F}_t\}_{t\geq 0},\mathcal{P})$ and a Hilbert space $\mathcal{H}$ with the norm $\|\cdot\|_{\mathcal{H}}$, define the Hilbert space

$L^2_{\mathcal{F}}(0,T;\mathcal{H}):=\{\phi(\cdot,\cdot):[0,T]\times\Omega\rightarrow\mathcal{H}|\phi(\cdot,\cdot)$ is an $\mathcal{F}_t-$ adapted, $\mathcal{H}-$ valued measurable process on $[0,T]$ and $\mathbf{E}\int_0^T\|\phi(t,\omega)\|^2dt<\infty\}$,

With the norm

$$\|\phi(\cdot)\|_{\mathcal{F},2}=\left(\mathbf{E}\int_0^T\|\phi(t,\omega)\|^2_{\mathcal{H}}dt\right)^{\frac{1}{2}}.$$

Consider the following linear stochastic differential equation (SDE) subject to Markovian jumps defined by

$$\begin{cases}dx(t)=[A(t,r_t)x(t)+B(t,r_t)u(t)]dt+\\ \qquad\qquad [C(t,r_t)x(t)+D(t,r_t)u(t)]dw(t), t\in[0,T],\\ x(s)=y,\end{cases} \tag{3.1.3}$$

where $(s,y)\in[0,T]\times\mathbb{R}^n$ are the initial time and initial state, respectively, and an admissible control $u(\cdot)$ is an $\mathcal{F}_t$-adapted, $\mathbb{R}^{n_u}$-valued measurable process on $[0,T]$. The set of all admissible controls is denoted by $\mathcal{U}_{ad}\equiv L^2_{\mathcal{F}}(0,T;\mathbb{R}^{n_u})$. The solution $x(\cdot)$ of the Eq. (3.1.3) is called the response of the control $u(\cdot)\in\mathcal{U}_{ad}$, and $(x(\cdot),u(\cdot))$ is called an admissible pair. Here, $w(t)$ is a one-dimensional standard $\mathcal{F}_t-$ Brownian motion on $[0,T]$ (with $w(0)=0$). r_t is a Markov chain adapted to

$\mathcal{F}_t$, taking values in $\Xi = \{1, \cdots, l\}$, with the transition probabilities specified by (3.1.2). In addition, we assume that the processes r_t and $w(t)$ are independent.

For each (s, y) and $u(\cdot) \in \mathcal{U}_{ad}$, the associated cost is

$$J(s,y,i;u(\cdot)) = \mathbf{E}\left\{\int_s^T \begin{bmatrix} x(t) \\ u(t) \end{bmatrix}' \begin{bmatrix} Q(t,r_t) & L(t,r_t) \\ L'(t,r_t) & R(t,r_t) \end{bmatrix} \begin{bmatrix} x(t) \\ u(t) \end{bmatrix} dt + x'(T)H(r_T)x(T) | r_s = i\right\}. \tag{3.1.4}$$

In (3.1.3) and (3.1.4), $A(t, r_t) = A_i(t)$, etc. whenever $r_t = i$, and $H(r_T) = H_i$ whenever $r_T = i$, whereas $A_i(\cdot)$ etc. are given matrix-valued functions and H_i are given matrices, $i = 1, \ldots, l$. The objective of the optimal control problem is to minimize the cost function $J(s, y, i; u(\cdot))$, for a given $(s, y) \in [0, T) \times \mathbb{R}^n$, over all $u(\cdot) \in \mathcal{U}_{ad}$. The value function is defined as

$$V(s,y,i) = \inf_{u(\cdot)\in\mathcal{U}_{ad}} J(s,y,i;u(\cdot)). \tag{3.1.5}$$

Definition 3.1.1 The optimization problem (3.1.3)–(3.1.5) is called well-posed if

$$V(s,y,i) \geq -\infty, \forall (s,y) \in [0,T) \times \mathbb{R}^n, \ \forall i = 1, \ldots .l.$$

An admissible pair $(x^*(\cdot), u^*(\cdot))$ is called optimal (with respect to the initial condition (s, y, i)) if $u^*(\cdot)$ achieves the infimum of $J(s, y, i; u(\cdot))$.

The following basic assumption will be in force throughout this section.

Assumption 3.1.1 The data appearing in the LQ problem (3.1.3)–(3.1.5) satisfy, for every i,

$$\begin{cases} A_i(\cdot), C_i(\cdot) \in L^\infty(0,T;\mathbb{R}^{n\times n}), \\ B_i(\cdot), D_i(\cdot) \in L^\infty(0,T;\mathbb{R}^{n\times n_u}), \\ Q_i(\cdot) \in L^\infty(0,T;\mathcal{S}^n), \\ L_i(\cdot) \in L^\infty(0,T;\mathbb{R}^{n\times n_u}), \\ R_i(\cdot) \in L^\infty(0,T;\mathcal{S}^{n_u}), \\ H_i \in \mathcal{S}^n. \end{cases}$$

We emphasize again that we are dealing with an indefinite LQ problem, namely, $Q_i(t)$, $R_i(t)$ and H_i are all possibly indefinite.

Lemma 3.1.1 [1] *Let a matrix $M \in \mathbb{R}^{m\times n}$ be given. Then there exists a unique matrix $M^\dagger \in \mathbb{R}^{n\times m}$ such that*

$$\begin{cases} M^\dagger M M^\dagger = M^\dagger, & MM^\dagger M = M, \\ (M^\dagger M)' = M^\dagger M, & (MM^\dagger)' = MM^\dagger, \end{cases}$$

where the matrix $M^\dagger$ is called the Moore–Penrose pseudo inverse of M.

Now we introduce a new type of coupled differential Riccati equations associated with the LQ problem (3.1.3)–(3.1.5).

Definition 3.1.2 The following system of constrained differential equations (with the time argument t suppressed)

$$\begin{cases} \dot{P}_i + P_iA_i + A_i'P_i + C_i'P_iC_i + Q_i + \sum_{j=1}^{l} \pi_{ij}P_j \\ -(P_iB_i + C_i'P_iD_i + L_i)(R_i + D_i'P_iD_i)^\dagger(B_i'P_i + D_i'P_iC_i + L_i') = 0, \\ P_i(T) = H_i, \\ (R_i + D_i'P_iD_i)(R_i + D_i'P_iD_i)^\dagger(B_i'P_i + D_i'P_iC_i + L_i') \\ -(B_i'P_i + D_i'P_iC_i + L_i') = 0, \\ R_i + D_i'P_iD_i \geq 0, \quad a.e.\ t \in [0, T], \quad i = 1, \ldots, l \end{cases} \tag{3.1.6}$$

is called a system of coupled generalized (differential) Riccati equations (CGREs).

Lemma 3.1.2 (generalized Itô's formula) [2] *Let $x(t)$ satisfy*

$$dx(t) = b(t, x(t), r_t)dt + \sigma(t, x(t), r_t)dw(t).$$

And $\varphi(\cdot, \cdot, i) = C^2([0, T] \times \mathbb{R}^n), i = 1, \ldots, l$, be given. Then,

$$d\varphi(t, x(t), r_t) = \Gamma\varphi(t, x(t), r_t)dt + \varphi_x'(t, x(t), r_t)\sigma(t, x(t), r_t)dw(t),$$

where

$$\begin{aligned} \Gamma\varphi(t, x, i) = {} & \varphi_t(t, x, i) + b'(t, x, i)\varphi_x(t, x, i) \\ & + \frac{1}{2}\mathrm{tr}[\sigma'(t, x, i)\varphi_{xx}(t, x, i)\sigma(t, x, i)] + \sum_{j=1}^{l} \pi_{ij}\varphi(t, x, j). \end{aligned}$$

Lemma 3.1.3 [3, 4] *For a symmetric matrix M, we have*

(i) $(M^\dagger)' = M^\dagger$;
(ii) $M^\dagger M = MM^\dagger$;
(iii) *$M \geq 0$ if and only if $M^\dagger \geq 0$.*

Lemma 3.1.4 (Extended Schur's lemma) [5] *Let matrix $M = M'$, N and $R = R'$ be given with appropriate dimensions. Then the following conditions are equivalent:*

(i) $M - NR^{\dagger}N' \geq 0$, *and* $N(I - RR^{\dagger}) = 0$, $R \geq 0$;

(ii) $\begin{bmatrix} M & N \\ N' & R \end{bmatrix} \geq 0$;

(iii) $\begin{bmatrix} R & N' \\ N & M \end{bmatrix} \geq 0$.

Lemma 3.1.5 [6] *Let matrices L, M, and N be given with appropriate sizes. Then the following matrix equation*

$$LXM = N \tag{3.1.7}$$

has a solution X if and only if

$$LL^{\dagger}XM^{\dagger}M = N. \tag{3.1.8}$$

Moreover, any solution to (3.1.7) *is represented by*

$$X = L^{\dagger}NM^{\dagger} + S - L^{\dagger}LNM^{\dagger}M, \tag{3.1.9}$$

where S is a matrix with an appropriate size.

3.1.1.2 Main Results

In this section, we will show that the solvability of the CGREs is sufficient for the well-posedness of the LQ problem and the existence of an optimal feedback control. In addition, all optimal controls can be obtained via the solution to the CGREs (3.1.6).

Theorem 3.1.1 *If the CGREs* (3.1.6) *admit a solution* $(P_1(\cdot), \cdots, P_l(\cdot)) \in \mathcal{C}^1(0, T; \mathcal{S}_l^n)$, *then the stochastic LQ problem* (3.1.3)–(3.1.5) *is well-posed. Moreover, the set of all optimal controls with respect to the initial* $(s, y) \in [0, T) \times \mathbb{R}^n$ *is determined by the following (parameterized by* (Y_i, z_i)*):*

$$\begin{aligned} u(t) = -\sum\nolimits_{i=1}^{l} \Big\{ \Big[& (R_i(t) + D_i(t)'P_i(t)D_i(t))^{\dagger}(B_i(t)'P_i(t) + D_i(t)'P_i(t)C_i(t) + L_i(t)') \\ & + Y_i(t) - (R_i(t) + D_i(t)'P_i(t)D_i(t))^{\dagger}(R_i(t) + D_i(t)'P_i(t)D_i(t))Y_i(t)\Big]x + z_i(t) \\ & - (R_i(t) + D_i(t)'P_i(t)D_i(t))^{\dagger}(R_i(t) + D_i(t)'P_i(t)D_i(t))z_i(t)\Big\}\chi_{r_t=i}(t), \end{aligned} \tag{3.1.10}$$

where $Y_i(\cdot) \in L^2_{\mathcal{F}}(0, T; \mathbb{R}^{n_u \times n})$ *and* $z_i(\cdot) \in L^2_{\mathcal{F}}(0, T; \mathbb{R}^{n_u})$. *Furthermore, the value function is uniquely determined by* $(P_1(\cdot), \ldots, P_l(\cdot)) \in \mathcal{C}^1(0, T; \mathcal{S}_l^n)$:

$$V(s,y,i) \equiv \inf_{u(\cdot)\in\mathcal{U}_{ad}} J(s,y,i;u(\cdot)) = y'P_i(s)y, \quad i=1,\ldots,l. \tag{3.1.11}$$

Proof Let $(P_1(\cdot),\ldots,P_l(\cdot)) \in C^1(0,T;\mathcal{S}_l^n)$ be a solution of the CGREs (3.1.6). Setting $\varphi(t,x,i)=x'P_i(t)x$ and applying the generalized Itô's formula (Lemma 3.1.2) to the linear system (3.1.3), we have

$$\begin{aligned}
&\mathbf{E}\left[x(T)'H_{r_T}x(T)\right] - y'P_i(s)y \\
&\quad= \mathbf{E}\left[x(T)'H_{r_T}x(T) - x(s)'P(r_s)x(s)|r_s=i\right] \\
&\quad= \mathbf{E}[\varphi(T,x(T),r_T) - \varphi(s,x(s),r_s)|r_s=i] \\
&\quad= \mathbf{E}\left\{\int_s^T \Gamma\varphi(t,x(t),r_t)|r_s=i\right\},
\end{aligned}$$

where

$$\begin{aligned}
\Gamma\varphi(t,x,i) &= \varphi_t(t,x,i) + b'(t,x,i)\varphi_x(t,x,i) \\
&\quad+\frac{1}{2}\mathrm{tr}[\sigma'(t,x,i)\varphi_{xx}(t,x,i)\sigma(t,x,i)] + \sum_{j=1}^{l}\pi_{ij}\varphi(t,x,j) \\
&= x'\left[\dot{P}_i(t) + P_i(t)A_i(t) + A_i(t)'P_i(t) + C_i(t)'P_i(t)C_i(t) + \sum\nolimits_{j=1}^{l}\pi_{ij}P_j(t)\right]x \\
&\quad+2u'\left[B_i(t)'P_i(t) + D_i(t)'P_i(t)C_i(t)\right]x + u'D_i(t)'P_i(t)D_i(t)u.
\end{aligned}$$

Hence, we can express the cost function as follows

$$\begin{aligned}
&J(s,y,i;u(\cdot)) \\
&\quad= y'P_i(s)y + \mathbf{E}\left\{\int_s^T [\Gamma\varphi(t,x(t),r_t) + x(t)'Q(t,r_t)x(t)\right. \\
&\qquad\left.+2u(t)'L(t,r_t)'x(t) + u(t)'R(t,r_t)u(t)]dt|r_s=i\right\}.
\end{aligned} \tag{3.1.12}$$

From the definition of the CGREs, we have

$$\begin{aligned}
&\Gamma\varphi(t,x,i) + x'Q_i(t)x + 2uL_i(t)'x + u'R_i(t)u \\
&\quad= x'\left[\dot{P}_i(t) + P_i(t)A_i(t) + A_i(t)'P_i(t) + C_i(t)'P_i(t)C_i(t) + Q_i(t) + \sum\nolimits_{j=1}^{l}\pi_{ij}P_j(t)\right]x \\
&\qquad+2u'\left[B_i(t)'P_i(t) + D_i(t)'P_i(t)C_i(t) + L_i(t)'\right]x + u'\left[R_i(t) + D_i(t)'P_i(t)D_i(t)\right]u.
\end{aligned}$$

Now, let $Y_i(\cdot) \in L^2_{\mathcal{F}}(0,T;\mathbb{R}^{n_u\times n})$ and $z_i(\cdot) \in L^2_{\mathcal{F}}(0,T;\mathbb{R}^{n_u})$ be given for every i. Set

$$G_i^1(t) = Y_i(t) - \left[R_i(t) + D_i(t)'P_i(t)D_i(t)\right]^{\dagger}\left[R_i(t) + D_i(t)'P_i(t)D_i(t)\right]Y_i(t),$$
$$G_i^2(t) = z_i(t) - \left[R_i(t) + D_i(t)'P_i(t)D_i(t)\right]^{\dagger}\left[R_i(t) + D_i(t)'P_i(t)D_i(t)\right]z_i(t).$$

Applying Proposition 3.1.3 and Lemma 3.1.4-(ii), we have for $k = 1, 2$,

$$\left[R_i(t) + D_i(t)'P_i(t)D_i(t)\right]G_i^k(t) = \left[R_i(t) + D_i(t)'P_i(t)D_i(t)\right]^{\dagger}G_i^k(t) = 0, \quad (3.1.13)$$

And

$$[P_i(t)B_i(t) + C_i(t)P_i(t)D_i(t) + L_i(t)]G_i^k(t) = 0.$$

Hence

$$\begin{aligned}
&\Gamma\varphi(t, x, i) + x'Q_i(t)x + 2uL_i(t)'x + u'R_i(t)u \\
&= \left[u + (G_i^1(t) - K_i(t))x + G_i^2(t)\right]'\left[R_i(t) + D_i(t)'P_i(t)D_i(t)\right] \\
&\quad \times \left[u + (G_i^1(t) - K_i(t))x + G_i^2(t)\right],
\end{aligned}$$

where

$$K_i(t) = -\left[R_i(t) + D_i(t)'P_i(t)D_i(t)\right]^{\dagger}\left[B_i(t)'P_i(t) + D_i(t)'P_i(t)C_i(t) + L_i(t)'\right].$$

Then the Eq. (3.1.4) can be expressed as

$$\begin{aligned}
&J(s, y, i; u(\cdot)) \\
&= y'P_i(s)y + E\left\{\int_s^T \left[u(t) + (G_i^1(t, r_t) - K_i(t, r_t))x + G_i^2(t, r_t)\right]'\right. \\
&\quad \times \left[R_i(t, r_t) + D_i(t, r_t)'P_i(t, r_t)D_i(t, r_t)\right] \\
&\quad \times \left[u(t) + (G_i^1(t, r_t) - K_i(t, r_t))x + G_i^2(t, r_t)\right]dt | r_s = i\},
\end{aligned} \quad (3.1.14)$$

where $P(t, r_t) = P_i(t)$, $K(t, r_t) = K_i(t)$ and $G^k(t, r_t) = G_i^k(t)$ whenever $r_t = i$, $k = 1, 2$. Thus, $J(s, y, i; u(\cdot))$ is minimized by the control given by (3.1.10) with the optimal value being $y'P_i(s)y$.

Theorem 3.1.1 presents a sufficient condition for the existence of optimal control, now let's explore its necessary condition.

Theorem 3.1.2 *Assume that $Q_i(t)$ and $R_i(t)$ are continuous in t for every i. In addition, assume that the LQ problem* (3.1.3)–(3.1.5) *is well-posed and a given feedback control $\bar{u}(t) = \sum_{i=1}^{l} \bar{K}_i(t)x(t)\chi_{r_t=i}(t)$ is optimal for* (3.1.3)–(3.1.5) *with respect to any initial $(s, y) \in [0, T) \times \mathbb{R}^n$. Then the CGREs* (3.1.6) *must have a solution $(P_1(\cdot), \ldots, P_l(\cdot)) \in C^1(0, T; S_l^n)$. Moreover, the optimal feedback control $\bar{u}(t) = \sum_{i=1}^{l} \bar{K}_i(t)x(t)\chi_{r_t=i}(t)$ can be represented via* (3.1.10) with $z(t) \equiv 0$.

Proof By the dynamic programming approach, the value functions $V(s,y,i)$ satisfy the following HJB equations for $i = 1,\ldots,l$

$$
\begin{aligned}
&V_s(s,y,i) + \min_u\{y'Q_iy + 2y'L_iu + u'R_iu + [A_iy + B_iu]'V_y(s,y,i) \\
&\quad + \frac{1}{2}[C_iy + D_iu]'V_{yy}(s,y,i)[C_iy + D_iu] + \sum\nolimits_{j=1}^{l}\pi_{ij}V(s,y,i)\Big\} = 0,
\end{aligned} \tag{3.1.15}
$$

with the boundary condition

$$
V(T,y,i) = y'H_iy. \tag{3.1.16}
$$

In view of the assumption of the theorem, a candidate value function can be represented as

$$
V(s,y,i) = y'P_i(s)y, \quad i = 1,\ldots,l, \tag{3.1.17}
$$

For a matrix $P_i(\cdot) \in \mathcal{S}^m$, suppose that $P_i(t)$ is differentiable at any $t \in [0,T]$. Substituting (3.1.17) into (3.1.15), we have the equations (s is suppressed)

$$
\begin{cases}
y'(\dot{P}_i + P_iA_i + A_i'P_i + C_i'P_iC_i + Q_i + \sum_{j=1}^{l}\pi_{ij}P_j)y \\
\quad + \min\limits_u\{u'(R_i + D_i'P_iD_i)u + 2y'(P_iB_i + C_i'P_iD_i + L_i)u\} = 0, \\
P_i(T) = H_i \qquad i = 1,\ldots,l.
\end{cases} \tag{3.1.18}
$$

By assumption, a minimizer u in (3.1.18) is given by $u(s,y,i) = K_i(s)y$ for i, and hence (3.1.18) are reduced to the following equations,

$$
\begin{cases}
y'(\dot{P}_i + P_iA_i + A_i'P_i + C_i'P_iC_i + Q_i + \sum_{j=1}^{l}\pi_{ij}P_j)y \\
\quad + \min\limits_{K_i}\{y'[K_i'(R_i + D_i'P_iD_i)K_i + 2(P_iB_i + C_i'P_iD_i + L_i)K_i]y\} = 0, \\
P_i(T) = H_i \qquad i = 1,\ldots,l.
\end{cases} \tag{3.1.19}
$$

The second term of the left-hand side of the first equation above reaches the minimum if and only if

$$
\frac{\partial}{\partial K_i}[K_i'(R_i + D_i'P_iD_i)K_i + 2(P_iB_i + C_i'P_iD_i + L_i)K_i]\big|_{K_i=\bar{K}_i} = 0, \ i = 1,\ldots,l.
$$

i.e.

$$(R_i + D_i'P_iD_i)\bar{K}_i + (B_i'P_i + D_i'P_iC_i + L_i') = 0, \quad i = 1, \ldots, l. \tag{3.1.20}$$

Setting $L_i = R_i + D_i'P_iD_i$, $M_i = I$, and $N_i = -B_i'P_i + D_i'P_iC_i + L_i'$, $i = 1, \ldots, l$, by applying Lemma 3.1.5 to the Eq. (3.1.20), we have

$$(R_i + D_i'P_iD_i)(R_i + D_i'P_iD_i)'(B_i'P_i + D_i'P_iC_i + L_i') = B_i'P_i + D_i'P_iC_i + L_i'.$$

First of all, by virtue of the assumption we know a priori that the Eq. (3.1.20) do have a solution $\bar{K}_i$, and $\bar{K}_i$ has the following form

$$\begin{aligned}\bar{K}_i = - & \Big[(R_i + D_i'P_iD_i)^{\dagger}(B_i'P_i + D_i'P_iC_i + L_i') \\ & + Y_i - (R_i + D_i'P_iD_i)^{\dagger}(R_i + D_i'P_iD_i)Y_i\Big], \quad i = 1, \ldots, l.\end{aligned} \tag{3.1.21}$$

Replacing $(\bar{K}_1(\cdot), \ldots, \bar{K}_l(\cdot))$ into the first l equations of (3.1.19), we can see by a simple calculation that $(P_1(\cdot), \ldots, P_l(\cdot)) \in C^1(0, T; \mathcal{S}_l^n)$ satisfies the following equations

$$\begin{aligned}&\dot{P}_i + P_iA_i + A_i'P_i + C_i'P_iC_i + Q_i + \sum\nolimits_{j=1}^{l} \pi_{ij}P_j \\ &\quad - (P_iB_i + C_i'P_iD_i + L_i)(R_i + D_i'P_iD_i)^{\dagger}(B_i'P_i + D_i'P_iC_i + L_i') = 0, \quad i = 1, \ldots, l.\end{aligned} \tag{3.1.22}$$

So we easily conclude that $(P_1(\cdot), \ldots, P_l(\cdot)) \in C^1(0, T; \mathcal{S}_l^n)$ solves (3.1.6). The representation of $(\bar{K}_1(\cdot), \ldots, \bar{K}_l(\cdot))$ is given by (3.2.21). This completes the proof.

3.1.2 *Infinite-Time Horizon Case*

3.1.2.1 Problem Formulation

To facilitate the narrative, first define the following space

$L_2^{loc}(\mathbb{R}^{n_u}) := \{\phi(\cdot,\cdot) : [0, T] \times \Omega \to \mathcal{H} | \phi(\cdot,\cdot)$ is $\mathcal{F}_t$-adapted, Lebesgue measurable, and $\mathbf{E}\int_0^T \|\phi(t,\omega)\|^2 dt < +\infty, \ \forall T \geq 0\}$.

Consider the linear stochastic differential equation subject to Markovian jumps defined by

$$\begin{cases} dx(t) = [A(r_t)x(t) + B(r_t)u(t)]dt + [C(r_t)x(t) + D(r_t)u(t)]dw(t), \\ x(0) = x_0 \in \mathbb{R}^n, \end{cases} \tag{3.1.23}$$

Where $A(r_t) = A_i, B(r_t) = B_i$, $C(r_t) = C_i$ and $D(r_t) = D_i$ when $r_t = i$, while $A_i(\cdot)$, etc., $i = 1, \cdots, l$, are given matrices of suitable sizes. A process $u(\cdot)$ is called a control if $u(\cdot) \in L_2^{loc}(\mathbb{R}^{n_u})$.

Definition 3.1.3 A control $u(\cdot)$ is called (mean-square) stabilizing with respect to (w.r.t.) a given initial state (x_0, i) if the corresponding state $x(\cdot)$ of (3.1.23) with $x(0) = x_0$ and $r_0 = i$ satisfies $\lim_{t\to\infty} \mathbf{E}\|x(t)\|^2 = 0$.

Definition 3.1.4 The system (3.1.23) is called (mean-square) stabilizable if there exists a feedback control $u^*(t) = \sum_{i=1}^{l} K_i \chi_{r_t=i}(t)x(t)$, where $K_1, \ldots, K_l$ are given matrices, which is stabilizing w.r.t. any initial state (x_0, i).

Next, for a given $(x_0, i) \in \mathbb{R}^n \times \{1, 2, \ldots, l\}$, we define the corresponding set of admissible controls:

$$\mathcal{U}(x_0, i) = \left\{u(\cdot) \in L^2_{\mathcal{F}}(\mathbb{R}^{n_u}) | u(\cdot) \text{ is mean-square stabilizing w.r.t.} (x_0, i)\right\}.$$

Where the integer n_u is the dimension of the control variable. It is easily seen that $\mathcal{U}(x_0, i)$ is a convex subset of $L_2^{loc}(\mathbb{R}^{n_u})$.

For each $(x_0, i, u(\cdot)) \in \mathbb{R}^n \times \{1, 2, \ldots, l\} \times \mathcal{U}(x_0, i)$, the optimal control problem is to find a control which minimizes the following quadratic cost associated with (3.1.23)

$$J(x_0, i; u(\cdot)) = E\left\{\int_0^{+\infty} \begin{bmatrix} x(t) \\ u(t) \end{bmatrix}' \begin{bmatrix} Q(r_t) & L(r_t) \\ L(r_t)' & R(r_t) \end{bmatrix} \begin{bmatrix} x(t) \\ u(t) \end{bmatrix} dt | r_0 = i\right\}, \qquad (3.1.24)$$

where $Q(r_t) = Q_i, R(r_t) = R_i$ and $L(r_t) = L_i$ when $r_t = i$, while Q_i, etc., $i = 1, \ldots, l$, are given matrices with suitable sizes. The value function V is defined as

$$V(x_0, i) = \inf_{u(\cdot) \in \mathcal{U}(x_0, i)} J(x_0, i; u(\cdot)). \qquad (3.1.25)$$

Definition 3.1.5 The LQ problem (3.1.23)–(3.1.25) is called well-posed if

$$-\infty < V(x_0, i) < +\infty, \quad \forall (x_0, i) \in \mathbb{R}^n \times \Xi.$$

A well-posed problem is called attainable (w.r.t. (x_0, i)) if there is a control $u^*(\cdot) \in \mathcal{U}(x_0, i)$ that achieves $V(x_0, i)$. In this case the control $u^*(\cdot)$ is called optimal (w.r.t. (x_0, i)).

Assumption 3.1.2 The system (3.1.23) is mean-square stabilizable.

Mean-square stabilizability is a standard assumption in an infinite-horizon LQ control problem. In words, it basically ensures that there is at least one meaningful control, in the sense that the corresponding state trajectory is square integrable (hence does not "blow up"), with respect to any initial conditions. The problem would be trivial without this assumption.

Assumption 3.3.3 The data appearing in the LQ problem (3.1.23)–(3.1.25) satisfy, for every i,

$$A_i, C_i \in \mathbb{R}^{n\times n}, B_i, D_i \in \mathbb{R}^{n\times n_u}, Q_i \in \mathcal{S}^n, L_i \in \mathbb{R}^{n\times n_u}, R_i \in \mathcal{S}^{n_u}.$$

3.1.2.2 Main Results

Theorem 3.1.3 *For the LQ problem* (3.1.23)–(3.1.25), *if the following ARE* (3.1.26) *has a maximal solution* $X(i) \geq 0$, $i = 1,\ldots,l$,

$$\begin{aligned} &X(i)A(i) + A'(i)X(i) + C'(i)X(i)C(i) - X(i)B(i)R^{-1}(i)B'(i)X(i) \\ &\quad + Q(i) + \sum_{j=1}^{l} \pi_{ij}X(j) = 0. \end{aligned} \tag{3.1.26}$$

Then the optimal feedback law is

$$u(t) = \sum_{i=1}^{l} K(i)\chi_{r_t=i}(t)x(t) = -\sum_{i=1}^{l} R^{-1}(i)B'(i)X(i)\chi_{r_t=i}(t)x(t). \tag{3.1.27}$$

And $J(x_0, i; u) \geq \mathbf{E}[x'(0)X^*(i)x(0)]$. Furthermore, if ARE (3.1.26) has a solution, then the solution is the maximal solution $X(i) = P^*(i)$, and $P(i) = P^*(i)$ is the solution to the following semi-definite dynamic programming

$$\text{maxmize } \mathbf{Tr}(P(i)), \tag{3.1.28a}$$

s.t.

$$\begin{bmatrix} X(i)A(i) + A'(i)X(i) + C'(i)X(i)C(i) & \\ + Q(i) + \sum_{j=1}^{l} \pi_{ij}X(j) & P(i)B(i) \\ B'(i)P(i) & R(i) \end{bmatrix} \geq 0. \tag{3.1.28b}$$

Proof The proof is similar with LQ problem in a finite time horizon, here we omitted it.

3.2 Stochastic Nash Differential Games with Two Player

3.2.1 Finite-Time Horizon Case

3.2.1.1 Problem Formulation

First, we consider a stochastic Nash differential game with two player on a finite horizon $[0, T]$, N-player case is similar.

Consider the following Markov jump linear systems described by stochastic differential equation

$$\begin{cases} dx(t) = [A(t,r_t)x(t) + B_1(t,r_t)u(t) + B_2(t,r_t)v(t)]dt \\ \qquad\qquad + [C(t,r_t)x(t)\,, + D_1(t,r_t)u(t) + D_2(t,r_t)v(t)]dw(t), \\ x(s) = y \in \mathbb{R}^n, \end{cases} \tag{3.2.1}$$

Where $(s,y) \in [0,T) \times \mathbb{R}^n$ are the initial time and initial state, respectively, and two admissible controls $u(\cdot)$ and $v(\cdot)$ are $\mathcal{F}_t$-adapted, $\mathbb{R}^{n_u}$- and $\mathbb{R}^{n_v}$-valued measurable process on $[0, T]$. The sets of all admissible controls are denoted by $\mathcal{U} \equiv L^2_{\mathcal{F}}(0,T;\mathbb{R}^{n_u})$ and $\mathcal{V} \equiv L^2_{\mathcal{F}}(0,T;\mathbb{R}^{n_v})$.

For each (s,y) and $(u(\cdot),v(\cdot)) \in \mathcal{U} \times \mathcal{V}$, the cost functional $J_k(s,y,i;u(\cdot),v(\cdot))$ for player k is

$$J_k(s,y,i;u(\cdot),v(\cdot)) = \mathbf{E}\left\{\int_s^T z'(t)M_k(t,r_t)z(t)dt + x'(T)H_k(r_T)x(T)|r_s = i\right\},$$
$$z(t) = \begin{bmatrix} x(t) \\ u(t) \\ v(t) \end{bmatrix},\ M_k(t,r_t) = \begin{bmatrix} Q_k(t,r_t) & L_{k1}(t,r_t) & L_{k2}(t,r_t) \\ L'_{k1}(t,r_t) & R_{k1}(t,r_t) & 0 \\ L'_{k2}(t,r_t) & 0 & R_{k2}(t,r_t) \end{bmatrix}, k = 1,2. \tag{3.2.2}$$

In (3.2.1) and (3.2.2), $A(t,r_t) = A_i(t)$, etc. whenever $r_t = i$, and $H_k(r_T) = H_{ki}$, $k = 1$, 2, whenever $r_T = i$, whereas $A_i(\cdot)$ etc. are given matrix-valued functions and H_{ki} are given matrices, $i = 1,\ldots,l$.

Assumption 3.2.1 The data appearing in the finite horizon stochastic Nash differential game problem (3.2.1)–(3.2.2) satisfy, for every i,

$$\begin{cases} A_i(\cdot), C_i(\cdot) \in L^\infty(0,T;\mathbb{R}^n), & B_{1i}(\cdot), D_{1i}(\cdot) \in L^\infty(0,T;\mathbb{R}^{n\times n_u}), \\ B_{2i}(\cdot), D_{2i}(\cdot) \in L^\infty(0,T;\mathbb{R}^{n\times n_v}), & Q_{1i}(\cdot) \in L^\infty(0,T;\mathcal{S}^n), \\ Q_{2i}(\cdot) \in L^\infty(0,T;\mathcal{S}^n), & R_{11i}(\cdot) \in L^\infty(0,T;\mathcal{S}^{n_u}), \\ L_{11i}(\cdot) \in L^\infty(0,T;\mathbb{R}^{n\times n_u}), & L_{12i}(\cdot) \in L^\infty(0,T;\mathbb{R}^{n\times n_v}), \\ L_{21i}(\cdot) \in L^\infty(0,T;\mathbb{R}^{n\times n_u}), & L_{22i}(\cdot) \in L^\infty(0,T;\mathbb{R}^{n\times n_v}), \\ R_{22i}(\cdot) \in L^\infty(0,T;\mathcal{S}^{n_v}), & H_{1i} \in \mathcal{S}^n, \quad H_{2i} \in \mathcal{S}^n. \end{cases}$$

Now, let's give the form definition of finite time stochastic Nash differential games:

Definition 3.2.1 For each $(s, y) \in [0, T) \times \mathbb{R}^n$, finding an admissible control pair $(u^*(\cdot), v^*(\cdot)) \in \mathcal{U} \times \mathcal{V}$, such that

$$\begin{cases} J_1(s, y, i; u^*(\cdot), v^*(\cdot)) \leq J_1(s, y, i; u(\cdot), v^*(\cdot)), & \forall u(\cdot) \in \mathcal{U}, \\ J_2(s, y, i; u^*(\cdot), v^*(\cdot)) \leq J_2(s, y, i; u^*(\cdot), v(\cdot)), & \forall v(\cdot) \in \mathcal{V}. \end{cases} \tag{3.2.3}$$

The strategy pair $(u^*(\cdot), v^*(\cdot))$ which satisfying (3.2.3) is called the Nash equilibrium of the game.

3.2.1.2 Main Results

With the help of the relevant conclusions of differential game with one-person discussed in 3.1, it is easy to obtain the following conclusions.

Theorem 3.2.1 *For the finite time stochastic Nash differential game* (3.2.1)–(3.2.2), *there exists the Nash equilibrium* $(u^*(\cdot), v^*(\cdot))$, *if and only if the following coupled generalized differential Riccati equations (with time t supressed)*

$$\begin{cases} \dot{P}_{1i} + P_{1i}\bar{A}_i + \bar{A}_i'P_{1i} + \bar{C}_i'P_{1i}\bar{C}_i + \bar{Q}_{1i} + \sum_{j=1}^{l} \pi_{ij}P_{1j} - (P_{1i}B_{1i} + \bar{C}_i'P_{1i}D_{1i} + L_{11i}) \\ \times (R_{11i} + D_{1i}'P_{1i}D_{1i})^{-1}(B_{1i}'P_{1i} + D_{1i}'P_{1i}\bar{C}_i + L_{11i}') = 0, \\ P_{1i}(T) = H_{1i}, \\ R_{11i} + D_{1i}'P_{1i}D_{1i} > 0, i = 1, \ldots, l. \end{cases} \tag{3.2.4a}$$

$$K_{1i} = -(R_{11i} + D_{1i}'P_{1i}D_{1i})^{-1}(B_{1i}'P_{1i} + D_{1i}'P_{1i}\bar{C}_i + L_{11i}'), \tag{3.2.4b}$$

$$\begin{cases} \dot{P}_{2j} + P_{2j}\tilde{A}_j + \tilde{A}_j'P_{2j} + \tilde{C}_j'P_{2j}\tilde{C}_j + \tilde{Q}_{2j} + \sum_{k=1}^{l} \pi_{jk}P_{2k} - \left(P_{2j}B_{2j} + \tilde{C}_j'P_{2j}D_{2j} + L_{22j}\right) \\ \times \left(R_{22j} + D_{2j}'P_{2j}D_{2j}\right)^{-1}\left(B_{2j}'P_{2j} + D_{2j}'P_{2j}\tilde{C}_j + L_{22j}'\right) = 0, \\ P_{2j}(T) = H_{2j}, \\ R_{22j}(j) + D_{2j}'(j)P_{2j}(j)D_{2j}(j) > 0\ ,\ j = 1, \ldots, l. \end{cases} \tag{3.2.4c}$$

$$K_{2j} = -\left(R_{22j} + D_{2j}'P_{2j}D_{2j}\right)^{-1}\left(B_{2j}'P_{2j} + D_{2j}'P_{2j}\tilde{C}_j + L_{22j}'\right). \tag{3.2.4d}$$

where

$$\bar{A}_i = A_i + B_{2i}K_{2i}, \bar{C}_i = C_i + D_{2i}K_{2i}, \bar{Q}_{1i} = Q_{1i} + L_{12i}K_{2i} + K_{2i}'L_{12i}' + K_{2i}'R_{12i}K_{2i},$$
$$\tilde{A}_j = A_j + B_{1j}K_{1j}, \tilde{C}_j = C_j + D_{1j}K_{1j}, \tilde{Q}_{2j} = Q_{2j} + L_{21j}K_{1j} + K_{1j}'L_{21j}' + K_{1j}'R_{21j}K_{1j}.$$

admit a solution $P(\cdot) = (P_1(\cdot), P_2(\cdot)) \in \mathcal{C}^1(0, T; \mathcal{S}_l^n) \geq 0,$ *where* $P_1(\cdot) = (P_{11}(\cdot), \ldots, P_{1l}(\cdot))$, $P_2(\cdot) = (P_{21}(\cdot), \ldots, P_{2l}(\cdot))$.

Denote $F_{1i}^*(t) = K_{1i}(t)$, $F_{2i}^*(t) = K_{2i}(t)$, then the Nash equilibrium strategy $(u^*(\cdot), v^*(\cdot))$ can be represented by

$$u^*(t) = \sum_{i=1}^{l} F_{1i}^*(t)\chi_{r_t=i}(t)x(t), v^*(t) = \sum_{i=1}^{l} F_{2i}^*(t)\chi_{r_t=i}(t)x(t).$$

Moreover, the optimal value is

$$J_k(s, y, i; u^*(\cdot), v^*(\cdot)) = y'P_{ki}(s)y, \quad k = 1, 2.$$

Proof These results can be proved by using the concept of Nash equilibrium described in Definition 3.2.1 as follows. Given $v^*(t) = \sum_{i=1}^{l} F_{2i}^*(t)\chi_{r_t=i}(t)x(t)$ is the optimal control strategy implemented by player P_2, player P_1 facing the following optimization problems:

$$\min_{u(\cdot)\in\mathcal{U}} \mathbf{E}\left\{\int_s^T \begin{bmatrix} x(t) \\ u(t) \end{bmatrix}' \begin{bmatrix} \bar{Q}_1(t, r_t) & L_{11}(t, r_t) \\ L_{11}'(t, r_t) & R_{11}(t, r_t) \end{bmatrix} \begin{bmatrix} x(t) \\ u(t) \end{bmatrix} dt + x'(T)H_1(r_T)x(T)|r_s = i\right\},$$

s.t.

$$\begin{cases} dx(t) = [\bar{A}(t, r_t)x(t) + B_1(t, r_t)u(t)]dt + [\bar{C}(t, r_t)x(t) + D_1(t, r_t)u(t)]dW(t), \\ x(s) = y. \end{cases} \tag{3.2.5}$$

where

$$\bar{Q}_1 = Q_1 + (F_2^*)'L_{12}' + L_{12}F_2^* + (F_2^*)'R_{12}F_2^*.$$

Note that the above optimization problem defined in (3.2.5) is a standard stochastic LQ problem. Applying Theorem 3.1.1 to this optimization problem as

$$\begin{bmatrix} \bar{Q}_1(r_t) & L_{11}(r_t) \\ L_{11}'(r_t) & R_{11}(r_t) \end{bmatrix} \Rightarrow \begin{bmatrix} Q_1 & L_1 \\ L_1' & R_{11} \end{bmatrix}, \bar{A} \Rightarrow A, \bar{C} \Rightarrow C.$$

We can easily get the optimal control and the optimal value function

$$u^*(t) = \sum_{i=1}^{l} F_{1i}^*(t)\chi_{r_t=i}(t)x(t),\ J_1(s,y,i;u^*(\cdot),v^*(\cdot)) = y'P_{1i}(s)y,\ i=1,\ldots,l. \tag{3.2.6}$$

Similarly, we can prove that $v^*(t) = \sum_{i=1}^{l} F_{2i}^*(t)\chi_{r_t=i}(t)x(t)$ is the optimal control strategy of player P_2.

This completes the proof of Theorem 3.2.1.

3.2.2 *Infinite-Time Horizon Case*

3.2.2.1 Problem Formulation

In this subsection, we discuss the stochastic Nash differential games on time interval $[0,\infty)$. Before giving the problem to be discussed, first define the following space

$L_2^{loc}(\mathbb{R}^m) := \{\phi(\cdot,\cdot) : [0,\infty)\times\Omega \to \mathbb{R}^m | \phi(\cdot,\cdot)$ is $\mathcal{F}_t$-adapted, Lebesgue measurable, and $\mathbf{E}\int_0^T \|\phi(t,\omega)\|^2 dt < \infty,\ \forall T>0\}$.

Consider the following Markov jump linear systems defined by

$$\begin{cases} dx(t) = [A(r_t)x(t) + B_1(r_t)u(t) + B_2(r_t)v(t)]dt \\ \qquad\qquad + [C(r_t)x(t) + D_1(r_t)u(t) + D_2(r_t)v(t)]dw(t), \\ x(0) = x_0. \end{cases} \tag{3.2.7}$$

where $A(r_t) = A(i)$, $B_1(r_t) = B_1(i)$, $B_2(r_t) = B_2(i)$, $C(r_t) = C(i)$, $D_1(r_t) = D_1(i)$ and $D_2(r_t) = D_2(i)$, when $r_t = i$, $i = 1,\ldots,l$, while $A(i)$, etc., are given matrices of suitable sizes. $u(\cdot) \in \mathcal{U} \equiv L_2^{loc}(\mathbb{R}^{n_u})$ and $v(\cdot) \in \mathcal{V} \equiv L_2^{loc}(\mathbb{R}^{n_v})$ are two admissible control processes, which represents the control strategies of these two players.

Definition 3.2.2 [7] The stochastically controlled system described by Itô's equation $dx(t) = [A(r_t)x(t) + B(r_t)u(t)]dt + [C(r_t)x(t) + D(r_t)u(t)]dw(t)$, $x(0) = x_0$ is mean-square stabilizable, if there exists a feedback control $u(t) = \sum_{i=1}^{l} K(i)\chi_{r_t=i}(t)x(t)$, where $K(1),\ldots,K(l)$ are given matrices, which is stabilizing w.r.t. any initial value $x(0) = x_0$, $r_0 = i$, the closed-loop system

$$dx(t) = [A(r_t) + B(r_t)K(r_t)]x(t)dt + [C(r_t) + D(r_t)K(r_t)]dw(t)$$

is asymptotically mean-square stable, i.e., $\lim_{t\to\infty} \mathbf{E}\left[\|x(t)\|^2 | r_0 = i\right] = 0$.

Next, for a given $(x_0, i) \in \mathbb{R}^n \times \Xi$, we define the corresponding sets of admissible controls:

$\bar{\mathcal{U}}(x_0, i) = \{(u(\cdot), v(\cdot)) \in \mathcal{U} \times \mathcal{V} | (u(\cdot), v(\cdot))$ is mean-square stabilizing w.r.t. $(x_0, i)\}$.

For each (x_0, i) and $(u(\cdot), v(\cdot)) \in \bar{\mathcal{U}}(x_0, i)$, the cost function $J_k(x_0, i; u(\cdot), v(\cdot))$ is

$$J_k(x_0, i; u(\cdot), v(\cdot)) = \mathbf{E}\left\{\int_0^{\infty} z'(t) M_k(r_t) z(t) dt | r_0 = i\right\},$$

$$z(t) = \begin{bmatrix} x(t) \\ u(t) \\ v(t) \end{bmatrix}, \; M_k(r_t) = \begin{bmatrix} Q_k(r_t) & L_{k1}(r_t) & L_{k2}(r_t) \\ L'_{k1}(r_t) & R_{k1}(r_t) & 0 \\ L'_{k2}(r_t) & 0 & R_{k2}(r_t) \end{bmatrix}, \; k = 1, 2. \tag{3.2.8}$$

In (3.2.7) and (3.2.8), $A(r_t) = A(i), \ldots,$ when $r_t = i$, while $A(i)$, etc., are given matrices with suitable sizes.

The form definition of infinite-time horizon stochastic Nash differential game is given below:

Definition 3.2.3 For each $(x_0, i) \in \mathbb{R}^n \times \Xi$, finding an admissible control pair $(u^*(\cdot), v^*(\cdot)) \in \bar{\mathcal{U}}(x_0, i)$, such that

$$\begin{cases} J_1(s, y, i; u^*(\cdot), v^*(\cdot)) \leq J_1(s, y, i; u(\cdot), v^*(\cdot)), & \forall u(\cdot) \in \mathcal{U}, \\ J_2(s, y, i; u^*(\cdot), v^*(\cdot)) \leq J_2(s, y, i; u^*(\cdot), v(\cdot)), & \forall v(\cdot) \in \mathcal{V}. \end{cases} \tag{3.2.9}$$

the strategy pair $(u^*(\cdot), v^*(\cdot))$ which satisfying (3.2.9) is called the Nash equilibrium of the game.

3.2.2.2 Main Results

Assumption 3.2.2 The system (3.2.7) is mean-square stabilizable.

Similar to the finite-time horizon stochastic Nash games discussed in last subsection, we can get the corresponding results of the infinite-time horizon stochastic Nash games stated as Theorem 3.2.2, which can be verified by following the line of Theorem 3.2.1.

Theorem 3.2.2 *Suppose Assumption* 3.2.1 *holds, the infinite-time horizon stochastic Nash differential game* (3.2.7)–(3.2.8) *has a Nash equilibrium* $(u^*(\cdot), v^*(\cdot))$, *if and only if the following algebraic Riccati equations admit a solution* $P = (P_1, P_2) \in \mathcal{S}_l^n \times \mathcal{S}_l^n \geq 0$ *with* $P_1 = (P_1(1), \ldots, P_1(l))$, $P_2 = (P_2(1), \ldots, P_2(l))$:

$$\begin{cases} P_1(i)\bar{A}(i)+\bar{A}'(i)P_1(i)+\bar{C}_1'(i)P_1(i)\bar{C}_1(i)+\bar{Q}_1(i)+\sum_{j=1}^{l}\pi_{ij}P_1(j) \\ -\left(P_1(i)B_1(i)+\bar{C}_1'(i)P_1(i)D_1(i)+L_{11}(i)\right)\left(R_{11}(i)+D_1'(i)P_1(i)D_1(i)\right)^{-1} \\ \times\left(B_1'(i)P_1(i)+D_1'(i)P_1(i)\bar{C}_1(i)+L_{11}'(i)\right)=0, \\ R_{11}(i)+D_1'(i)P_1(i)D_1(i)>0,\ i\in\Xi. \end{cases} \tag{3.2.10a}$$

$$K_1=-\left(R_{11}(i)+D_1'(i)P_1(i)D_1(i)\right)^{-1}\left(B_1'(i)P_1(i)+D_1'(i)P_1(i)\bar{C}_1(i)+L_{11}'(i)\right), \tag{3.2.10b}$$

$$\begin{cases} P_2(j)\tilde{A}(j)+\tilde{A}'(j)P_2(j)+\tilde{C}_2'(j)P_2(j)\tilde{C}_2(j)+\tilde{Q}_2(j)+\sum_{k=1}^{l}\pi_{jk}P_2(k) \\ -\left(P_2(j)B_2(j)+\tilde{C}_2'(j)P_2(j)D_2(j)+L_{22}(j)\right)\left(R_{22}(j)+D_2'(j)P_2(j)D_2(j)\right)^{-1} \\ \times\left(B_2'(j)P_2(j)+D_2'(j)P_2(j)\tilde{C}_2(j)+L_{22}'(j)\right)=0, \\ R_{22}(j)+D_2'(j)P_2(j)D_2(j)>0,\ j\in\Xi. \end{cases} \tag{3.2.10c}$$

$$K_2=-\left(R_{22}(j)+D_2'(j)P_2(j)D_2(j)\right)^{-1}\left(B_2'(j)P_2(j)+D_2'(j)P_2(j)\tilde{C}_2(j)+L_{22}'(j)\right). \tag{3.2.10d}$$

where

$$\bar{A}=A+B_2K_2,\bar{C}_1=C+D_2K_2,\bar{Q}_1=Q_1+L_{12}K_2+K_2'L_{12}'+K_2'R_{12}K_2,$$
$$\tilde{A}=A+B_1K_1,\tilde{C}_2=C+D_1K_1,\tilde{Q}_2=Q_2+L_{21}K_1+K_1'L_{21}'+K_1'R_{21}K_1$$

The equilibrium strategies and optimal cost function are

$$u^*(t)=\sum_{i=1}^{l}K_1(i)\chi_{r_t=i}(t)x(t),\ v^*(t)=\sum_{i=1}^{l}K_2(i)\chi_{r_t=i}(t)x(t).$$
$$J_k(x_0,i;u^*(\cdot),v^*(\cdot))=x_0'P_k(i)x_0,\quad k=1,\ 2,\quad i=1,\ldots,l.$$

3.2.3 Two Person Zero-Sum Stochastic Differential Game

In two person stochastic Nash differential games, when the sum of the two players' cost function is zero, i.e., $J_1=-J_2$, the game is degenerated to two person zero-sum stochastic differential game problem. Two person zero-sum stochastic differential game has been widely used in economics and management field, and this subsection is devoted to the theoretical study of this game.

3.2.3.1 Finite-Time Horizon Case

Consider the games described by the following linear stochastic differential equation with Markovian parameter jumps

$$\begin{cases} dx(t) = [A(t,r_t)x(t) + B_1(t,r_t)u(t) + B_2(t,r_t)v(t)]dt + \\ \qquad [D_0(t,r_t)x(t) + D_1(t,r_t)u(t) + D_2(t,r_t)v(t)]dw(t), \ t \in [0,T], \\ x(s) = y. \end{cases} \quad (3.2.11)$$

where $(s,y) \in [0,T) \times \mathbb{R}^n$ is the initial time and state, $u(\cdot)$ and $v(\cdot)$ are two admissible control processes, i.e., $\mathcal{F}_t$-adapted, $\mathbb{R}^{n_u}$-and $\mathbb{R}^{n_v}$-valued measurable process on $[0,T]$. The sets of all admissible controls are denoted by $\mathcal{U} \equiv L^2_{\mathcal{F}}(0,T;\mathbb{R}^{n_u})$ and $\mathcal{V} \equiv L^2_{\mathcal{F}}(0,T;\mathbb{R}^{n_v})$; $\{r_t\}$ and $\{w(t)\}$ are a Markov process and a standard one-dimensional Brownian motion which are independent.

For each (s,y) and $(u(\cdot),v(\cdot)) \in \mathcal{U} \times \mathcal{V}$, the cost function of the game is defined by:

$$J_\gamma(s,y,i;u(\cdot),v(\cdot)) = \mathbf{E}\left\{\int_s^T [x'(t)Q(t,r_t)x(t) + u'(t)R(t,r_t)u(t) - \gamma^2 v'(t)v(t)]dt \right. \\ \left. + x'(T)H(r_T)x(T)|r_s = i\right\}, \quad (3.2.12)$$

where $\gamma > 0$ is a given constant.

In (3.2.11) and (3.2.12), $A(t,r_t) = A_i(t),\ldots,$ whenever $r_t = i$, moreover, when $r_T = i$, $H(r_T) = H_i$. Referring to stochastic LQ problem, the corresponding value function is defined by:

$$V(s,y,i) = \inf_{u\in\mathcal{U}} \sup_{v\in\mathcal{V}} J_\gamma(s,y,i;u(\cdot),v(\cdot)) = \sup_{v\in\mathcal{V}} \inf_{u\in\mathcal{U}} J_\gamma(s,y,i;u(\cdot),v(\cdot)) \\ = J_\gamma(s,y,i;u^*(\cdot),v^*(\cdot)).$$

The problem is to look for $(u^*(\cdot),v^*(\cdot)) \in \mathcal{U} \times \mathcal{V}$ which is called the saddle point equilibrium for the game, such that for each $(s,y) \in [0,T) \times \mathbb{R}^n$. and $i \in \Xi$

$$J_\gamma(s,y,i;u^*(\cdot),v(\cdot)) \le J_\gamma(s,y,i;u^*(\cdot),v^*(\cdot)) \le J_\gamma(s,y,i;u(\cdot),v^*(\cdot)), \quad (3.2.13)$$

for each $(s,y) \in [0,T) \times \mathbb{R}^n$, and $i \in \Xi$.

Definition 3.2.4 The stochastic differential games (3.2.11)–(3.2.12) are well posed if

$$V(s,y,i) \ge -\infty, \forall (s,y) \in [0,T] \times \mathbb{R}^n, \ i \in \Xi.$$

Assumption 3.2.3 The data appearing in the game problem (3.2.11)–(3.2.12) satisfy, for every i,

$$\begin{cases} A_i(\cdot), D_{0i}(\cdot) \in L^\infty(0,T;\mathbb{R}^n), & B_{1i}(\cdot), D_{1i}(\cdot) \in L^\infty(0,T;\mathbb{R}^{n\times n_u}), \\ B_{2i}(\cdot), D_{2i}(\cdot) \in L^\infty(0,T;\mathbb{R}^{n\times n_v}), & Q_i(\cdot) \in L^\infty(0,T;\mathcal{S}^n), \\ R_i(\cdot) \in L^\infty(0,T;\mathcal{S}^{n_u}), & H_i \in \mathcal{S}^n. \end{cases}$$

Next, we will give the explicit form of the saddle point equilibrium strategy and the optimal cost function of the game combined with "completion square method". Since the control weighting matrix $R_i(\cdot)$ in the cost function may be indefinite, we replace inverse matrix with Moore-Penrose pseudo inverse matrix, and accompanied with constrained generalized differential Riccati equation. The main results are presented by Theorems 3.2.3 and 3.2.4.

Theorem 3.2.3 *For the Markov jump linear system* (3.2.11), *if the following generalized differential Riccati equation* (*with time t suprresed*)

$$\begin{cases} \dot{P}_i + \mathbf{N}(P_i) - \mathbf{S}'(P_i)\mathbf{R}^\dagger(P_i)\mathbf{S}(P_i) = 0, \\ P_i(T) = H_i, \\ R_i + \pi_{11}(P_i) > 0, -\gamma^2 I + \pi_{22}(P_i) > 0, \ a.e.\ t \in [0,T], \quad i = 1,\ldots,l. \end{cases} \tag{3.2.14}$$

admits a solution $(P_1(\cdot),\ldots,P_l(\cdot)) \in C^1(0,T;\mathcal{S}_l^n)$, *where*

$$\begin{cases} \mathbf{N}(P_i) = P_iA_i + A_i'P_i + \pi_{00}(P_i) + Q_i + \sum_{j=1}^{l} \pi_{ij}P_j, \\ \mathbf{S}(P_i) = \begin{bmatrix} B_{1i}'P_i + \pi_{10}(P_i) \\ B_{2i}'P_i + \pi_{20}(P_i) \end{bmatrix}, \\ \mathbf{R}(P_i) = \begin{bmatrix} R_i + \pi_{11}(P_i) & \pi_{12}(P_i) \\ \pi_{21}(P_i) & -\gamma^2 I + \pi_{22}(P_i) \end{bmatrix}, \\ \pi_{\tau\varsigma}(P_i) = D_{\tau i}'P_iD_{\varsigma i}, \tau, \varsigma = 0,1,2, \end{cases} \tag{3.2.15}$$

then the game (3.2.11)–(3.2.12) *has a saddle point equilibrium* $\bar{u}^*(\cdot)$, *and for any initial value* $(s,y) \in [0,T) \times \mathbb{R}^n$, *its explicit expression is*

$$\bar{u}^*(t) = -\sum_{i=1}^{l} \mathbf{R}^\dagger(P_i(t))\mathbf{S}(P_i(t))\chi_{\{r_t=i\}}(t)x(t).$$

Meanwhile, the optimal value is $V(s,y,i) = J_\gamma(s,y,i;\bar{u}^*(\cdot)) = y'P_i(s)y$, $i = 1,\ldots,l$.

Proof Since the two person zero-sum differential game is a special case of two person nonzero-sum differential game, so the proof of Theorem 3.2.3 can similarly referring to Theorem 3.2.1, and we omitted here.

3.2.3.2 Infinite-Time Horizon Case

In this subsection, we consider the two person zero-sum stochastic differential games on time interval $[0, \infty)$. Firstly, we define the following space

$$L_2^{loc}(\mathbb{R}^m) := \{\phi(\cdot,\cdot) : [0,\infty) \times \Omega \to \mathbb{R}^m | \phi(\cdot,\cdot)$$

is $\mathcal{F}_t$-adapted, Lebesgue measurable, and $\mathbf{E}\int_0^T \|\phi(t,\omega)\|^2 dt < \infty,\ \forall T > 0\}$.

For notation's simplicity, considering the following controlled Markov jump linear systems:

$$\begin{cases} dx(t) = [A(r_t)x(t) + B_1(r_t)u(t) + B_2(r_t)v(t)]dt + [D_0(r_t)x(t) + D_1(r_t)u(t)]dw(t), \\ x(0) = x_0, \end{cases} \tag{3.2.16}$$

where $A(r_t) = A(i)$, $B_1(r_t) = B_1(i)$, $B_2(r_t) = B_2(i)$, $D_0(r_t) = D_0(i)$ and $D_1(r_t) = D_1(i)$, when $r_t = i$, $i = 1, \ldots, l$, while $A(i)$, etc., are given matrices of suitable sizes. $u(\cdot) \in \mathcal{U} \equiv L_2^{loc}(\mathbb{R}^{n_u})$ and $v(\cdot) \in \mathcal{V} \equiv L_2^{loc}(\mathbb{R}^{n_v})$ are two admissible control processes, which represents the control strategies of these two players.

For system (3.2.16) and $(x_0, i) \in \mathbb{R}^n \times \Xi$, the corresponding sets of admissible controls are denoted by:

$\bar{\mathcal{U}}(x_0, i) = \{(u(\cdot), v(\cdot)) \in \mathcal{U} \times \mathcal{V} | (u(\cdot), v(\cdot))$ is mean-square stabilizing w.r.t. $(x_0, i)\}$.

For each (x_0, i) and $(u(\cdot), v(\cdot)) \in \bar{\mathcal{U}}(x_0, i)$, the cost function is

$$J_\gamma(x_0, i; u(\cdot), v(\cdot)) = \mathbf{E}\left\{ \int_0^\infty (x'(t)Q(r_t)x(t) + u'(t)R(r_t)u(t) - \gamma^2 v'(t)v(t))dt | r_0 = i \right\}. \tag{3.2.17}$$

where $\gamma > 0$ is a given constant, and $Q(r_t) = Q(i)$, $R(r_t) = R(i)$, whenever $r_t = i$, $i = 1, \ldots, l$, while $Q(i)$, etc., are given matrices with suitable sizes. The value function is defined as

$$\begin{aligned} V(x_0, i) &= \inf_{u \in \mathcal{U}} \sup_{v \in \mathcal{V}} J_\gamma(x_0, i; u(\cdot), v(\cdot)) = \sup_{v \in \mathcal{V}} \inf_{u \in \mathcal{U}} J_\gamma(x_0, i; u(\cdot), v(\cdot)) \\ &= J_\gamma(x_0, i; u^*(\cdot), v^*(\cdot)). \end{aligned}$$

The problem is to look for $(u^*(\cdot), v^*(\cdot)) \in \bar{\mathcal{U}}(x_0, i)$ which is called the saddle point equilibrium for the game, such that

$$J_\gamma(s, y, i; u^*(\cdot), v(\cdot)) \le J_\gamma(s, y, i; u^*(\cdot), v^*(\cdot)) \le J_\gamma(s, y, i; u(\cdot), v^*(\cdot)), \quad i = 1, \ldots, l.$$

Definition 3.2.5 The stochastic differential games (3.2.16)–(3.2.17) are well posed if

$$-\infty < V(x_0, i) < +\infty, \quad \forall x_0 \in \mathbb{R}^n, \ \forall i = 1, \ldots, l.$$

Assumption 3.2.4 The system (3.2.16) is mean-square stabilizable.

Similar to the finite-time horizon stochastic Nash games discussed in last subsection, we can get the corresponding results of the infinite-time horizon stochastic Nash games stated as Theorem 3.2.3, which can be verified by following the line of Theorems 3.2.1 and 3.2.2.

Theorem 3.2.4 *Suppose Assumption* 3.2.4 *holds, for the Markov jump linear system* (3.2.16) *and* $(x_0, i) \in \mathbb{R}^n \times \Xi$, *the feedback control* $u^*(\cdot) = \sum_{i=1}^{l} K_1(i)\chi_{r_t=i}(t)x(t)$ *and* $v^*(\cdot) = \sum_{i=1}^{l} K_2(i)\chi_{r_t=i}(t)x(t)$ *is the equilibrium strategy of stochastic differential game* (3.2.16)–(3.2.17), *where* $K_1(i)$ *and* $K_2(i)$ *are given matrices with suitable size, if and only if the following algebraic Riccati equation*

$$\begin{cases} P(i)A(i) + A'(i)P(i) + D_0'(i)P(i)D_0(i) + Q(i) + \sum_{j=1}^{l} \pi_{ij}P(j) + \\ \gamma^{-2}P(i)B_2(i)B_2'(i)P(i) - [P(i)B_1(i) + D_0'(i)P(i)D_1(i)] \times \\ [R(i) + D_0'(i)P(i)D_0(i)]^{\dagger}[B_1'(i)P(i) + D_1'(i)P(i)D_0(i)] = 0, \\ R(i) + D_0'(i)P(i)D_0(i) > 0, \ i = 1, \ldots, l. \end{cases} \tag{3.2.18}$$

admits a solution $(P_1, \ldots, P_l) \in \mathcal{S}_l^n$. *In this case,*

$$K_1(i) = -[R(i) + D_0'(i)P(i)D_0(i)]^{\dagger}[B_1'(i)P(i) + D_1'(i)P(i)D_0(i)], K_2(i) \\ = \gamma^{-2}B_2'(i)P(i).$$

Meanwhile, the optimal value is $V(x_0, i) = J_\gamma(x_0, i; u^*(\cdot), v^*(\cdot)) = x_0'P(i)x_0$.

3.2.4 Numerical Example

In order to verify the correctness of the conclusions, consider all the coefficient matrices of the system (3.2.7) taking the following values:

$$\Xi = \{1,2\}, \Pi = \begin{bmatrix} -0.2 & 0.2 \\ 0.8 & -0.8 \end{bmatrix}, A(1) = \begin{bmatrix} 0 & 1 \\ -2 & -3 \end{bmatrix}, A(2) = \begin{bmatrix} 0 & 1 \\ 1 & 0 \end{bmatrix}, C(1) = \begin{bmatrix} 0.1 & 0 \\ 0 & 0.3 \end{bmatrix},$$
$$C(2) = \begin{bmatrix} 0.5 & 0 \\ 0 & 0.2 \end{bmatrix}, B_1(1) = \begin{bmatrix} 1 \\ 0 \end{bmatrix}, B_1(2) = \begin{bmatrix} 3 \\ 0 \end{bmatrix}, B_2(1) = \begin{bmatrix} 1 \\ 1 \end{bmatrix}, B_2(2) = \begin{bmatrix} 0 \\ 1 \end{bmatrix},$$
$$D_1(1) = \begin{bmatrix} 0 \\ 0.1 \end{bmatrix}, D_1(2) = \begin{bmatrix} 0 \\ 0 \end{bmatrix}, D_2(1) = \begin{bmatrix} 0 \\ 0.05 \end{bmatrix}, D_2(2) = \begin{bmatrix} 0 \\ 0.01 \end{bmatrix}, Q_1(1) = Q_1(2) = \begin{bmatrix} 1 & 0 \\ 0 & 2 \end{bmatrix},$$
$$Q_2(1) = Q_2(2) = \begin{bmatrix} 2 & 0 \\ 0 & 0.5 \end{bmatrix}, L_{11}(i) = L_{12}(i) = L_{21}(i) = L_{22}(i) = 0, \quad i = 1,2,$$
$$R_{11}(1) = R_{11}(2) = R_{22}(1) = R_{22}(2) = 0.1, R_{12}(1) = R_{12}(2) = R_{21}(1) = R_{21}(2) = 0.$$

Using the Newton's method proposed by Mukaidani [8–13] to solving (3.2.10a, 3.2.10b, 3.2.10c, 3.2.10d), we get the gain matrices $K_1(i)$ and $K_2(i)$ are

$$K_1(1) = [-2.979 \quad 0.986], K_1(2) = [-3.467 \quad -0.404];$$
$$K_2(1) = [-1.193 \quad -1.186], K_2(2) = [-0.241 \quad -4.311].$$

So the optimal control strategy of the system is

$$u^* = -2.979x_1 + 0.986x_2, v^* = -1.193x_1 - 1.186x_2, \text{ when } r_t = 1;$$
$$u^* = -3.467x_1 - 0.404x_2, v^* = -0.241x_1 - 4.311x_2, \text{ when } r_t = 2.$$

Under the control of $u^*(t)$ and $v^*(t)$, the system's Eq. (3.2.8) can be denoted as

$$dx(t) = A_C(r_t)x(t)dt + G(r_t)x(t)dw(t),$$

where

$$A_C(1) = \begin{bmatrix} -4.1720 & 0.8000 \\ -3.1930 & -4.1860 \end{bmatrix}, G(1) = \begin{bmatrix} 0.1000 & 0 \\ -0.3576 & 0.3393 \end{bmatrix};$$
$$A_C(2) = \begin{bmatrix} -10.4010 & -0.2120 \\ 0.7590 & -4.3110 \end{bmatrix}, G(2) = \begin{bmatrix} 0.5000 & 0 \\ -0.0024 & 0.1569 \end{bmatrix}.$$

Using Matlab with simulation step $\Delta = 0.001$, initial value $r_0 = 1$, $x_1(0) = 2$ and $x_2(0) = 1$, we obtain the state trajectories as shown in Figs. 3.1, 3.2 and 3.3.

As can be seen from Figs. 3.1, 3.2 and 3.3, under the control of $u^*(t)$ and $v^*(t)$, the closed-loop system is stable.

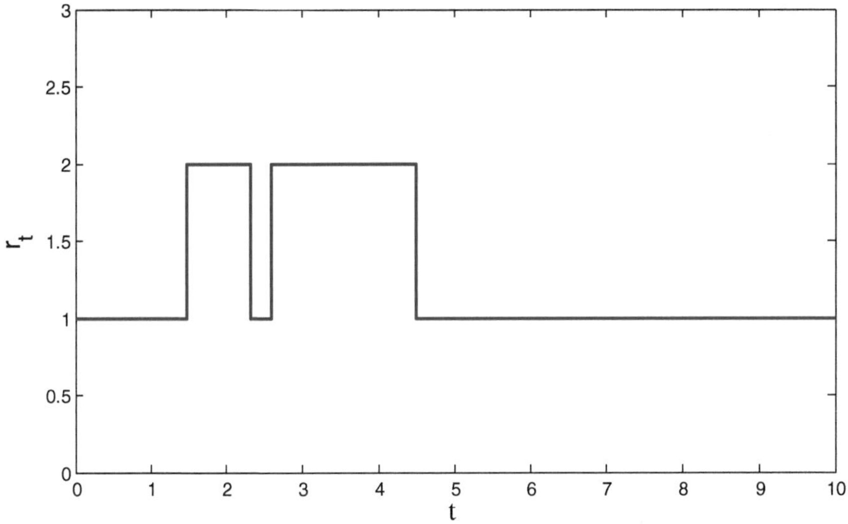

Fig. 3.1 Curve of r_t

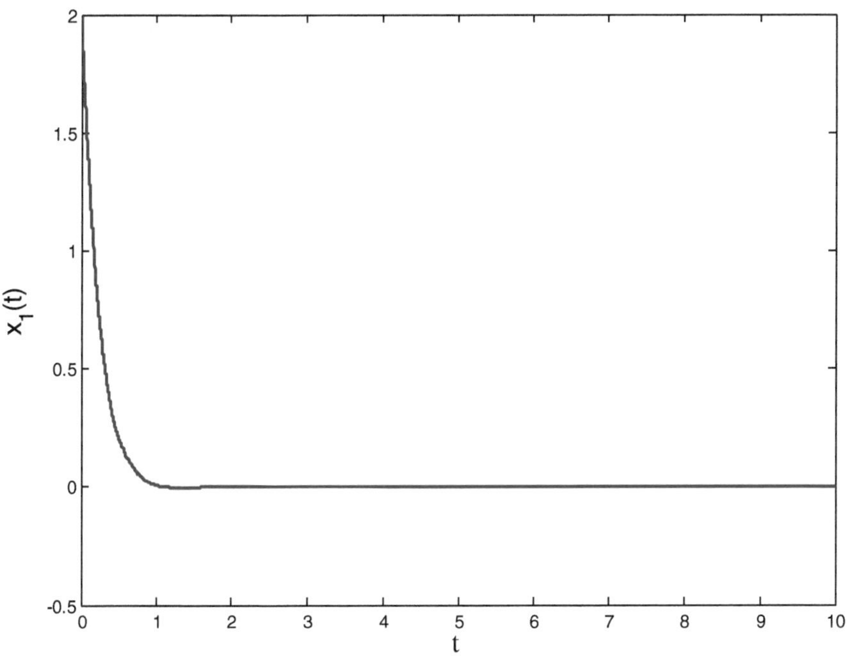

Fig. 3.2 Curve of $x_1(t)$

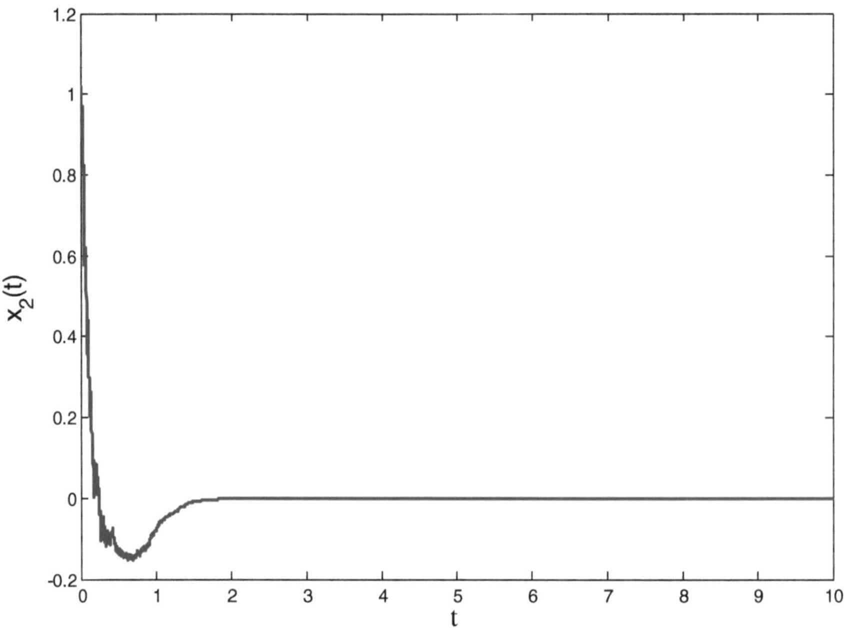

Fig. 3.3 Curve of $x_2(t)$

3.3 Stochastic Stackelberg Differential Game with Two Person

In this subsection, we will refer to Ref. [14], trying to extending the relevant conclusions from Itô stochastic system to Markov jump systems.

3.3.1 Problem Formulation

Consider the following game system with 2-players involving state-dependent noise.

$$\begin{cases} dx(t) = [A(r_t)x(t) + B_1(r_t)u_1(t) + B_2(r_t)u_2(t)]dt + C(r_t)x(t)dw(t), \\ x(0) = x_0. \end{cases} \tag{3.3.1}$$

where $x(t) \in \mathbb{R}^n$ represents the system state, $u_k(t) \in \mathbb{R}^{m_k}$, $k = 1, 2$ represent the k-th control inputs. It is assumed that the player denoted by u_2 is the leader and the player denoted by u_1 is the follower. In (3.3.1), $A(r_t) = A(i)$, $B_k(r_t) = B_k(i)$, $k = 1, 2$, $C(r_t) = C(i)$, when $r_t = i$, $i = 1, \ldots, l$, while $A(i)$, etc., are given matrices of suitable sizes.

Without loss of generality, the stochastic dynamic games are investigated under the following basic assumption:

Assumption 3.3.1 (A, B_k, C), $k = 1, 2$ is stabilizable.

For each initial value (x_0, i), the cost function for each strategy subset is defined by

$$J_k(x_0, i; u_1, u_2) = \mathbf{E}\left\{ \int_0^\infty \begin{bmatrix} x'(t)Q_k(r_t)x(t) + u_k'(t)R_{kk}(r_t)u_k(t) \\ + u_j'(t)R_{kj}(r_t)u_j(t) \end{bmatrix} dt | r_0 = i \right\}, \tag{3.3.2}$$

where $k = 1, 2$, $Q_k(r_t) = Q_k'(r_t) \geq 0$, $R_{kk}(r_t) = R_{kk}'(r_t) > 0$, $R_{kj}(r_t) = R_{kj}'(r_t) \geq 0,\ k \neq j$.

3.3.2 *Main Results*

Without loss of generality, we restrict the control strategy of each player as linear state feedback case, i.e., the closed-loop Stackelberg strategies $u_k(t) = u_k(x, t)$ have the following form

$$u_k(t) = \sum_{i=1}^{l} F_k(i)\chi_{r_t=i}(t)x(t).$$

The Stackelberg strategy of the game system (3.2.1)–(3.3.2) is defined as:

Definition 3.4.1 [14] a strategy set (u_1^*, u_2^*) is called a Stackelberg strategy if the following conditions hold

$$J_2(x_0, i; u_1^*, u_2^*) \leq J_2(x_0, i; u_1^o(u_2), u_2), \quad \forall u_2 \in \mathbb{R}^{m_2}, \tag{3.3.3}$$

where

$$J_1(x_0, i; u_1^o(u_2), u_2) = \min_v J_1(x_0, i; u_1, u_2), \tag{3.3.4}$$

and

$$u_1^* = u_1^0(u_2^*). \tag{3.3.5}$$

Theorem 3.3.1 *Suppose that the following cross-coupled algebraic matrix* Eqs. (3.3.6a–3.3.6e) *has solutions* $\bar{M}_k(i) \geq 0$, $\bar{N}_k(i)$, $k = 1, 2$ *and* $F_2(i)$

$$\begin{aligned}&A'_{F_1}(i)\bar{M}_1(i)+\bar{M}_1(i)A_{F_1}(i)+C'(i)\bar{M}_1(i)C(i)\\&-F'_1(i)R_{11}(i)F_1(i)+Q_{F_1}(i)+\sum_{j=1}^{l}\pi_{ij}\bar{M}_1(j)=0,\end{aligned}\tag{3.3.6a}$$

$$\begin{aligned}&A'_F(i)\bar{M}_2(i)+\bar{M}_2(i)A_F(i)+C'(i)\bar{M}_2(i)C(i)+Q_{F_2}(i)\\&+F'_1(i)R_{21}(i)F_1(i)+\sum_{j=1}^{l}\pi_{ij}\bar{M}_2(j)=0,\end{aligned}\tag{3.3.6b}$$

$$\begin{aligned}&A_{F_1}(i)\bar{N}_1(i)+\bar{N}_1(i)A'_{F_1}(i)+C(i)\bar{N}_1(i)C'(i)-B_1(i)R_{11}^{-1}(i)B'_1(i)\bar{M}_1(i)\bar{N}_1(i)\\&-\bar{N}_1(i)\bar{M}_1(i)B_1(i)R_{11}^{-1}(i)B'_1(i)+\pi_{ii}\bar{N}_1(i)-B_1(i)R_{11}^{-1}(i)B'_1(i)\bar{M}_2(i)\bar{N}_2(i)\\&-\bar{N}_2(i)\bar{M}_2(i)B_1(i)R_{11}^{-1}(i)B'_1(i)+B_1(i)R_{11}^{-1}(i)R_{21}(i)R_{11}^{-1}(i)B'_1(i)\bar{M}_1(i)\bar{N}_2(i)\\&+\bar{N}_2(i)\bar{M}_1(i)B_1(i)R_{11}^{-1}(i)R_{21}(i)R_{11}^{-1}(i)B'_1(i)=0,\end{aligned}\tag{3.3.6c}$$

$$A_F(i)\bar{N}_2(i)+\bar{N}_2(i)A'_F(i)+C(i)\bar{N}_2(i)C(i)+\pi_{ii}\bar{N}_2(i)+I_n=0,\tag{3.3.6d}$$

$$R_{12}(i)F_2(i)\bar{N}_1(i)+R_{22}(i)F_2(i)\bar{N}_2(i)+B'_2(i)(\bar{M}_1(i)\bar{N}_1(i)+\bar{M}_2(i)\bar{N}_2(i))=0,\tag{3.3.6e}$$

where

$$F_1(i)=-\sum_{i=1}^{l}R_{11}^{-1}(i)B'_1(i)\bar{M}_1(i),A_{F_1}(i)=A(i)+B_2(i)F_2(i),A_F(i)=A_{F_1}(i)+B_1(i)F_1(i),$$
$$Q_{F_1}(i)=Q_1(i)+F'_2(i)R_{12}(i)F_2(i),Q_{F_2}(i)=Q_2(i)+F'_2(i)R_{22}(i)F_2(i).$$

Denote $u_1^*(t)=\sum_{i=1}^{l}F_1(i)\chi_{r_t=i}(t)x(t)$ and $u_2^*(t)=\sum_{i=1}^{l}F_2(i)\chi_{r_t=i}(t)x(t)$, $i=1,\cdots,l$, then the strategy set (u_1^*,u_2^*) constitutes the Stackelberg strategy.

Proof Given arbitrary $u_2(t)=F_2(r_t)x(t)$, the corresponding u_1 is obtained by minimizing $J_1(x_0,i;u_1)$ with respect to u_1. Let us consider the minimizing problem for the closed-loop stochastic system with arbitrary strategies $u_2(t)=F_2(r_t)x(t)$

$$\begin{aligned}&\min_{u_1}\bar{J}_1(x_0,i;u_1)=\mathbf{E}\left\{\int_0^{\infty}\left[x'(t)Q_{F_1}(r_t)x(t)+u'_1(t)R_{11}(r_t)u_1(t)\right]dt|r_0=i\right\},\\&s.t.\\&dx(t)=[A_{F_2}(r_t)x(t)+B_1(r_t)u_1(t)]dt+C(r_t)x(t)dw(t).\end{aligned}\tag{3.3.7}$$

By using Theorem 3.1.2, the optimal state feedback controller $u_1^o(t)$ is given by

$$u_1^o(t) = \sum_{i=1}^{l} F_1(i)\chi_{r_t=i}(t)x(t) = -\sum_{i=1}^{l} R_{11}^{-1}(i)B_1'(i)\bar{M}_1(i)\chi_{r_t=i}(t)x(t), \tag{3.3.8}$$

where $\bar{M}_1(i)$ is the solution to

$$\begin{aligned} \mathsf{F}_1(\bar{M}_1(i), F_2(i)) &= A_{F_1}'(i)\bar{M}_1(i) + \bar{M}_1(i)A_{F_1}(i) + C'(i)\bar{M}_1(i)C(i) \\ &\quad - F_1'(i)R_{11}(i)F_1(i) + Q_{F_1}(i) + \sum_{j=1}^{l} \pi_{ij}\bar{M}_1(j) = 0. \end{aligned} \tag{3.3.9}$$

From (3.3.9) we can see that Eq. (3.3.6a) holds. On the other hand, if $A_F(i) = A_{F_1}(i) + B_1(i)F_1(i)$ is asymptotically mean square stable, then the cost J_2 of the leader can be represented as

$$\begin{aligned} &J_2(x_0, i; u_1^o(u_2), u_2) \\ &= J_2(x_0, i; F_1(r_t)x, F_2(r_t)x(t), = \mathbf{Tr}(\bar{M}_2(i)), \end{aligned} \tag{3.3.10}$$

where $\bar{M}_2(i)$ is the solution to

$$\begin{aligned} &\mathsf{F}_2(\bar{M}_1(i), \bar{M}_2(i), F_2(i)) \\ &= A_F'(i)\bar{M}_2(i) + \bar{M}_2(i)A_F(i) + C'(i)\bar{M}_2(i)C(i) + Q_{F_2}(i) \\ &\quad + F_1'(i)R_{21}(i)F_1(i) + \sum_{j=1}^{l} \pi_{ij}\bar{M}_2(j) = 0. \end{aligned} \tag{3.3.11}$$

From (3.3.11) we know (3.3.6b) holds. Let us consider the following Lagrangian $\mathbf{H}$

$$\begin{aligned} \mathbf{H}(\bar{M}_1(i), \bar{M}_2(i), F_2(i)) &= \mathbf{Tr}(\bar{M}_2(i)) + \mathbf{Tr}(\bar{N}_1(i)\mathsf{F}_1(\bar{M}_1(i), F_2(i))) \\ &\quad + \mathbf{Tr}(\bar{N}_2(i)\mathsf{F}_2(\bar{M}_1(i), \bar{M}_2(i), F_2(i))), \end{aligned} \tag{3.3.12}$$

where $\bar{N}_1(i)$ and $\bar{N}_2(i)$ are symmetric matrix of Lagrange multipliers.

As a necessary condition to minimization $\mathbf{Tr}(\bar{M}_2(i))$, we get

$$
\begin{aligned}
\frac{\partial \mathbf{H}}{\partial \bar{M}_1(i)} &= A_{F_1}(i)\bar{N}_1(i) + \bar{N}_1(i)A'_{F_1}(i) + C(i)\bar{N}_1(i)C'(i) \\
&\quad - B_1(i)R_{11}^{-1}(i)B'_1(i)\bar{M}_1(i)\bar{N}_1(i) - \bar{N}_1(i)\bar{M}_1(i)B_1(i)R_{11}^{-1}(i)B'_1(i) \\
&\quad + \pi_{ii}\bar{N}_1(i) - B_1(i)R_{11}^{-1}(i)B'_1(i)\bar{M}_2(i)\bar{N}_2(i) \\
&\quad - \bar{N}_2(i)\bar{M}_2(i)B_1(i)R_{11}^{-1}(i)B'_1(i) \\
&\quad + B_1(i)R_{11}^{-1}(i)R_{21}(i)R_{11}^{-1}(i)B'_1(i)\bar{M}_1(i)\bar{N}_2(i) \\
&\quad + \bar{N}_2(i)\bar{M}_1(i)B_1(i)R_{11}^{-1}(i)R_{21}(i)R_{11}^{-1}(i)B'_1(i) = 0,
\end{aligned} \tag{3.3.13a}
$$

$$
\begin{aligned}
\frac{\partial \mathbf{H}}{\partial \bar{M}_2(i)} &= A_F(i)\bar{N}_2(i) + \bar{N}_2(i)A'_F(i) + C(i)\bar{N}_2(i)C(i) \\
&\quad + \pi_{ii}\bar{N}_2(i) + I_n = 0,
\end{aligned} \tag{3.3.13b}
$$

$$
\begin{aligned}
\frac{1}{2}\frac{\partial \mathbf{H}}{\partial F_2(i)} &= R_{12}(i)F_2(i)\bar{N}_1(i) + R_{22}(i)F_2(i)\bar{N}_2(i) \\
&\quad + B'_2(i)(\bar{M}_1(i)\bar{N}_1(i) + \bar{M}_2(i)\bar{N}_2(i)) = 0.
\end{aligned} \tag{3.3.13c}
$$

Therefore, (3.3.6c)–(3.3.6e) hold. This completes the proof of Theorem 3.3.1.

3.4 Summary

For continuous-time Markov jump linear system, we firstly discussed the two person nonzero-sum stochastic differential game problem in finite-time horizon and infinite-time horizon. By using the related conclusion of stochastic LQ problem of Markov jump linear systems, we obtain the necessary and sufficient conditions for the existence of the system combined with Riccati equation method, which corresponds to the existence of the differential (algebraic) Riccati equation, and with the solution of Riccati equation, the optimal control strategy and explicit expression of the optimal value function of the system are given. Finally, numerical examples demonstrate the effectiveness of the obtained results. At the end, two person Stackelberg game problem of Markov jump linear systems in infinite-time horizon is discussed, and the existence condition of equilibrium strategy and explicit expression are given.

References

1. Penrose, R. (1955). A generalized inverse for matrices. In *Mathematical Proceedings of the Cambridge Philosophical Society* (Vol. 51, no. 3, pp. 406–413). Cambridge University Press.
2. Björk, T. (1980). Finite dimensional optimal filters for a class of Itô-processes with jumping parameters. *Stochastics, 4*(2), 167–183.
3. Rami, M. A., & Zhou, X. Y. (2000). Linear matrix inequalities, Riccati equations, indefinite stochastic linear quadratic controls. *IEEE Transactions on Automatic Control, 45*(6), 1131–1143.
4. Rami, M. A., Moore, J. B., & Zhou, X. Y. (2002). Indefinite stochastic linear quadratic control and generalized differential Riccati equation. *SIAM Journal on Control and Optimization, 40* (4), 1296–1311.
5. Chen, S., Li, X., & Zhou, X. Y. (1998). Stochastic linear quadratic regulators with indefinite control weight costs. *SIAM Journal on Control and Optimization, 36*(5), 1685–1702.
6. Li, X., Zhou, X. Y., & Rami, M. A. (2003). Indefinite stochastic linear quadratic control with markovian jumps in infinite time horizon. *Journal of Global Optimization, 27*(2–3), 149–175.
7. Mukaidani, H. (2006). A numerical analysis of the Nash strategy for weakly coupled large-scale systems. *IEEE Transactions on Automatic Control, 51*(8), 1371–1377.
8. Mukaidani, H. (2007). Newton's method for solving cross-coupled sign-indefinite algebraic Riccati equations for weakly coupled large-scale systems. *Applied Mathematics and Computation, 188*(1), 103–115.
9. Mukaidani, H. (2007). Numerical computation of sign indefinite linear quadratic differential games for weakly coupled large-scale systems. *International Journal of Control, 80*(1), 75–86.
10. Sagara, M., Mukaidani, H., & Yamamoto, T. (2008). Numerical solution of stochastic Nash games with state-dependent Noise for weakly coupled large-scale systems. *Applied Mathematics and Computation, 197*(2), 844–857.
11. Sagara, M., Mukaidani, H., & Yamamoto, T. (2008). Efficient numerical computations of soft constrained Nash strategy for weakly coupled large-scale systems. *Journal of Computers, 3*(9), 2–10.
12. Mukaidani, H. (2009). Soft-constrained stochastic Nash games for weakly coupled large-scale systems. *Automatica, 45*(5), 1272–1279.
13. Mukaidani, H. (2009). Robust guaranteed cost control for uncertain stochastic systems with multiple decision makers. *Automatica, 45*(7), 1758–1764.
14. Mukaidani, H. (2014). Stackelberg strategy for discrete-time stochastic system and its application to H_2/H_∞ control. In *American Control Conference* (ACC) (pp. 4488–4493). IEEE.

Chapter 4
Stochastic Differential Game of Discrete-Time Markov Jump Linear Systems

This chapter investigated the stochastic differential game theory of discrete-time markov jump linear systems, in which the state equation is described by Itô's stochastic algebraic equation. The Nash equilibrium strategies of the two person nonzero-sum differential game, the saddle point strategies of the two person zero-sum differential game, Stackelberg differential game were discussed in this chapter, and it was proved that sufficient conditions for the existence of the equilibrium strategy are equivalent to the solvability of the corresponding algebraic Riccati equations; moreover, the explicit solution of the optimal control strategy and the expression of the optimal value function were obtained. Finally, the numerical simulation examples were given.

4.1 Stochastic LQ Problem—Differential Game with One Person

4.1.1 Finite-Time Horizon

4.1.1.1 Problem Formulation

On a probabilistic space $(\Omega, \mathcal{F}, \{\mathcal{F}_t\}_{t\geq 0}, \mathcal{P})$, we consider a discrete-time markov jump linear systems of the following type:

$$\begin{cases} x(k+1) = A(r_k)x(k) + B(r_k)u(k) + A_1(r_k)x(k)w(k), \\ x(0) = x_0 \in \mathbb{R}^n, \end{cases} \tag{4.1.1}$$

where $x(k) \in \mathbb{R}^n$ is the state, $u(k) \in \mathbb{R}^m$ represent the system control inputs, x_0 is a deterministic vector. r_k Denotes a time-varying markov chain taking values in $\Xi = \{1, \ldots, l\}$ with transition probability matrix $\rho(k) = [p_{ij}(k)], p_{ij}(k) = \mathrm{P}(\theta(r+1) = j|r(k) = i)$. The coefficients $A(r_k), A_1(r_k), B(r_k)$ are assumed to be constant matrices

C.-k. Zhang et al., *Non-cooperative Stochastic Differential Game Theory of Generalized Markov Jump Linear Systems*, Studies in Systems, Decision and Control 67, DOI 10.1007/978-3-319-40587-2_4

with appropriate dimensions. $w(k)$ is a one-dimensional standard Brownian motion. Assuming that $w(k)$ is uncorrelated with $u(k)$, and is independent of the Markov chain r_k for $k \in N_T$. The initial value $r(0) = r_0$ is independent of the noise $w(k)$.

Let the optimal strategies for system (4.1.1) be given as

$$u(k) = K(r_k)x(k). \tag{4.1.2}$$

The purpose of the one person LQ differential games is to find the feedback controls with constant matrix gain $K(r_k)$ satisfying the following criterion:

$$J(u;x_0,i) = \mathbf{E}\left\{\left[x'(T)Mx(T) + \sum_{k=0}^{T-1}(x'(k)Q(r_k)x(k) + u'(k)R(r_k)u(k))\right]|r_0 = i\right\}, \tag{4.1.3}$$

which also minimizing

$$J(u^*;x_0,i) \le J(u;x_0,i), \tag{4.1.4}$$

where all the weighting matrices $R(r_k) \in \mathcal{S}_l^m, Q(r_k) \ge 0 \in \mathcal{S}_l^n$ in (4.1.1) and (4.1.2), when $r_k = i, i = 1,\ldots,l, A(r_k) = A(i)$, etc.

4.1.1.2 Main Results

The following theorem presents a sufficient condition for the existence of the optimal solutions to the finite-time stochastic LQ control problems.

Theorem 4.1.1 *For the system* (4.1.1) *with the criteria* (4.1.3), *if the following generalized algebraic Riccati equations admit a group of solutions* ($P^i(k)$; $K(i)$) *for any* (i, k)

$$\begin{cases} A'(i)\bar{P}^{(i)}(k+1)A(i) + A_1'(i)\bar{P}^{(i)}(k+1)A_1(i) + Q(i) \\ -P^{(i)}(k) + K'(i)B'(i)\bar{P}^{(i)}(k+1)A(i) = 0, \\ P(T) = M, B'(i)\bar{P}^{(i)}(k+1)B(i) + R(i) > 0, \\ K(i) = -\left(B'(i)\bar{P}^{(i)}(k+1)B(i) + R(i)\right)^{-1}B'(i)\bar{P}^{(i)}(k+1)A(i), \end{cases} \tag{4.1.5}$$

where $\{P^i(k) \in S_n\}$ *represents symmetric matrix indexed by the time k and the mode of operation i, and*

$$\bar{P}^{(i)}(k+1) = E(P_{r_{k+1}}(k+1)) = \sum_{j=1}^{l}\pi_{ij}P^{(j)}(k+1).$$

Then the finite time LQ stochastic games are solvable with $u^*(k) = K(i)\,x(k)$, *and the optimal cost functions incurred by playing strategies* $u^*(k)$ *are* $J(u^*(k)) = x_0'P^{(i)}(0)x_0, i \in \Xi$.

Proof If $P(\cdot) = (P^{(1)}(\cdot), P^{(2)}(\cdot), \ldots, P^{(l)}(\cdot)) \in \mathcal{S}_l^n$ is the solution of the Eq. (4.1.5), $x(\cdot)$ is the solution of the Eq. (4.1.3) corresponding to the admissible control $u(\cdot) \in \mathcal{U}[0,T]$. Considering the scalar function $Y(k,x) = x'(k)P^{(r_k)}(k)x(k)$, we can obtain

$$
\begin{aligned}
&\mathbf{E}\sum_{k=0}^{T-1}\left[\left(x(k+1)'P^{(r_{k+1})}(k+1)x(k+1) - x(k)'P^{(r_k)}(k)x(k)\right)|r_0 = i\right] \\
&= \mathbf{E}\left\{\sum_{k=0}^{T-1}\left[x'(k)\left(A'(i)\bar{P}^{(i)}(k+1)A(i) + A_1'(i)\bar{P}^{(i)}(k+1)A_1(i) - P^{(i)}(k)\right)x(k)\right.\right. \\
&\quad \left.\left. + 2u'(k)B'(i)\bar{P}^{(i)}(k+1)A(i)x(k) + u'(k)B'(i)\bar{P}^{(i)}(k+1)B(i)u(k)\right]\right\} \\
&= E[x'(T)P(T)x(T)] - x_0'P^{(i)}(0)x_0.
\end{aligned} \tag{4.1.6}
$$

Substituting (4.1.6) into (4.1.4), we have

$$
\begin{aligned}
J(u;x_0,i) &= -\mathbf{E}[x'(T)P(T)x(T)] + x_0'P^{(i)}(0)x_0 + \mathbf{E}[x'(T)Mx(T)] \\
&\quad + \mathbf{E}\left\{\sum_{k=0}^{T-1}\left[x'(k)\left(Q(i) + A_1'(i)\bar{P}^{(i)}(k+1)A_1(i) + A'(i)\bar{P}^{(i)}(k+1)A(i)\right)x(k)\right.\right. \\
&\quad \left.\left. + 2u'(k)B'(i)\bar{P}^{(i)}(k+1)A(i)x(k) + u'(k)\left(R(i) + B'(i)\bar{P}^{(i)}(k+1)B(i)\right)u(k)\right]\right\}.
\end{aligned} \tag{4.1.7}
$$

Using the completion square method to (4.1.7), we can obtain

$$
\begin{aligned}
J(u;x_0,i) &= -\mathbf{E}[x'(T)P(T)x(T)] + x_0'P^{(i)}(0)x_0 + \mathbf{E}[x'(T)Mx(T)] \\
&\quad + x'(k)\left[A'(i)\bar{P}^{(i)}(k+1)A(i) + A_1'(i)\bar{P}^{(i)}(k+1)A_1(i)\right. \\
&\quad \left. + Q(i) - P^{(i)}(k) + K'(i)B'(i)\bar{P}^{(i)}(k+1)A(i)\right]x(k) \\
&\quad + (u(k) - K(i)x(k))'\left(B'(i)\bar{P}^{(i)}(k+1)B(i) + R(i)\right)(u(k) - K(i)x(k)),
\end{aligned}
$$

where $K(i) = -\left(B'(i)\bar{P}^{(i)}(k+1)B(i) + R(i)\right)^{-1}B'(i)\bar{P}^{(i)}(k+1)A(i)$.

Hence, considering the Eq. (4.1.5), it is easy to deduce that

$$
J(u;x_0,i) \geq x_0'P^{(i)}(0)x_0 = J(u^*;x_0,i), i \in \Xi. \tag{4.1.8}
$$

this ends the proof. □

The sufficient conditions for the existence of the optimal control strategy are given in Theorem 4.1.1. The necessary conditions are given in the following Theorem 4.1.2.

Theorem 4.1.2 *For system (4.1.1), if* $u^*(k) = K(i)x(k)$ *is the optimal control strategy, where* $K(i) = -\left(B'(i)\bar{P}^{(i)}(k+1)B(i)+R(i)\right)^{-1}B'(i)\bar{P}^{(i)}(k+1)A(i), i \in \Xi$ *is a numerical matrix, then the algebraic Riccati Eq. (4.1.3) admits a solution* $P(\cdot) = (P^{(1)}(\cdot), P^{(2)}(\cdot), \ldots, P^{(l)}(\cdot)) \in S_l^n$ *and*

$$K(i) = -\left(B'(i)\bar{P}^{(i)}(k+1)B(i)+R(i)\right)^{-1}B'(i)\bar{P}^{(i)}(k+1)A(i).$$

Proof We can prove Theorem 4.1.2 by the method of dynamic programming. Reference [1] assuming that for any interval $kT < t < NT$, $k = 0, 1, \ldots, N-1$, select a quadratic value function as follows

$$\begin{aligned} &F_k(u; x(k), r_k) \\ &\quad = \mathbf{E}\left\{x'(T)Mx(T) + \sum_{j=k}^{T-1}\left[x'(j)Q(r_j)x(j) + u'(j)R(r_j)u(j)\right]\left|x(d), r_d, 0 \le d \le N\right.\right\}. \end{aligned}$$

Marking the minimization F_k over $u(k), u(k+1), \ldots, u(T-1)$ as $S(x(k), r_k, k)$. Given k, $x(k) = x^*(k), r_k = i$, then

$$\begin{aligned} S(x^*(k), i, k) &= \min_{u(k)}\{F_k\} \\ &= \min_{u(k)}[x'(k)Q(r_k)x(k) + u'(k)R(r_k)u(k)] + \min_{u(k)} E\{x'(T)Mx(T) \\ &\quad + \sum_{j=k+1}^{T-1}\left\{\left[\left(x'(j)Q(r_j)x(j) + u'(j)R(r_j)u(j)\right)\left|x(d), r_d, 0 \le d \le N\right.\right]\right\}\Bigg\} \\ &= \min_{u(k)}[x'(k)Q(r_k)x(k) + u'(k)R(r_k)u(k)] + E[S(x(k+1), r_{k+1}, k+1)|x^*(k), i]. \end{aligned} \tag{4.1.9}$$

Let $S(x(k), r_k, k) = x(k)'P^{(r_k)}(k)x(k)$, where $P^{(r_k)}(k)$ is a matrix to be determined, taking conditional expection on $S(x(k), r_k, k)$, we have

$$\begin{aligned} &\mathbf{E}[S(x(k+1), r_{k+1}, k+1)|x^*(k), i] \\ &= x^*(k)'\left(A'(i)\bar{P}^{(i)}(k+1)A(i) + A_1'(i)\bar{P}^{(i)}(k+1)A_1(i)\right)x^*(k) \qquad (4.1.10) \\ &\quad + 2u'(k)B'(i)\bar{P}^{(i)}(k+1)A(i)x^*(k) + u'(k)B'(i)\bar{P}^{(i)}(k+1)B(i)u(k). \end{aligned}$$

Substituting (4.1.10) into (4.1.9), removing the irrelevant items from $u(k)$, and calculating the first order condition, we have

$$u(k) = -\left(B'(i)\bar{P}^{(i)}(k+1)B(i)+R(i)\right)^{-1}B'(i)\bar{P}^{(i)}(k+1)A(i)x^*(k) = K(i)x^*(k). \tag{4.1.11}$$

According to the hypothesis of Theorem 4.1.2, we can know that the optimal control strategy of system (4.1.1) exists, so the $P^{(i)}(k)$ in (4.1.11) also exists.

Substituting (4.1.11) into (4.1.9), we have

$$\begin{aligned} S(x^*(k), i, k) = x^*(k)'\Big[&A'(i)\bar{P}^{(i)}(k+1)A(i)+A_1'(i)\bar{P}^{(i)}(k+1)A_1(i)+Q(i)\\ &+K'(i)B'(i)\bar{P}^{(i)}(k+1)A(i)\Big]x^*(k). \end{aligned}$$

According to the hypothesis, we can get $S(x^*(k), i, k) = x^*(k)'P^{(i)}(k)x^*(k)$, then

$$\begin{aligned} &A'(i)\bar{P}^{(i)}(k+1)A(i)+A_1'(i)\bar{P}^{(i)}(k+1)A_1(i)+Q(i)\\ &\quad+K'(i)B'(i)\bar{P}^{(i)}(k+1)A(i) - P^{(i)}(k) = 0. \end{aligned}$$

So we can know that the $P(\cdot) = (P^{(1)}(\cdot), P^{(2)}(\cdot), \cdots, P^{(l)}(\cdot)) \in \mathcal{S}_l^n$ is the solution of the Eq. (4.1.8).

This completes the proof. □

4.1.2 *Infinite-Time Horizon*

We still consider the discrete-time markov jump linear systems (4.1.1). For convenience of description we copy (4.1.1) as follows:

$$\begin{cases} x(k+1) = A(r_k)x(k) + B(r_k)u(k) + A_1(r_k)x(k)w(k),\\ x(0) = x_0 \in \mathbb{R}^n, \end{cases} \tag{4.1.12}$$

Definition 4.1.1 ([3]) The discrete time markov jump linear systems (4.1.12) is called mean-square stable if for any $(x_0, i) \in$ hollow body $R^n \times \Xi$, the corresponding state satisfies

$$\lim_{k\to\infty} \mathbf{E}\|x(k)\|^2 = 0.$$

Definition 4.1.2 ([3]) System (4.1.12) is called stabilizable in the mean square sense if there exists a feedback control $u^*(k) = \sum_{i=1}^{l} K(i)\chi_{r_k=i}(k)x(k)$ with $K(1),\cdots,K(l)$ are constant matrix, such that for any $(x_0, i) \in$ hollow body $R^n \times \Xi$, the following closed-loop system

$$\begin{cases} x(k+1) = A(r_k)x(k) + B(r_k)K(r_k)x(k) + A_1(r_k)x(k)w(k), \\ x(0) = x_0 \in \mathbb{R}^n, \end{cases}$$

is asymptotically mean square stable.

The purpose of the stochastic LQ problems is to find the feedback controls with constant matrix gain $K(r_k)$ satisfying the following criterion:

$$J(u; x_0, i) = \mathbf{E}\left\{\left[\sum_{k=0}^{\infty}(x'(k)Q(r_k)x(k) + u'(k)R(r_k)u(k))\right]|r_0 = i\right\}, \quad (4.1.13)$$

which also minimizing

$$J(u^*; x_0, i) \leq J(u; x_0, i), \quad (4.1.14)$$

where all the weighting matrices $R(r_k) \in \mathcal{S}_l^m, Q(r_k) \geq 0 \in \mathcal{S}_l^n$

In (4.1.12) and (4.1.13), when $r_k = i, i = 1,\ldots,l, A(r_k) = A(i)$, etc.

Assumption 4.1.1 *Systems (4.1.12) is mean-square stable.*

According to the relevant theory of stochastic optimal control, we can get Theorem 4.1.3, because the proof method is similar as the LQ problem in finite-time horizon, it is not repeated herein.

Theorem 4.1.3 *For the system* (4.1.12) *with the criteria* (4.1.13), *suppose the assumption* 4.1.1 *holds, if the following generalized algebraic Riccati equations admit a group of solutions* $P = (P^{(1)}, P^{(2)},\ldots,P^{(l)}) \in \mathcal{S}_l^n$ *for any (i, k)*

$$\begin{cases} A'(i)\bar{P}^{(i)}(k+1)A(i) + A_1'(i)\bar{P}^{(i)}(k+1)A_1(i) + Q(i) \\ -P^{(i)}(k) + K'(i)B'(i)\bar{P}^{(i)}(k+1)A(i) = 0, \\ B'(i)\bar{P}^{(i)}(k+1)B(i) + R(i) > 0, \\ K(i) = -\left(B'(i)\bar{P}^{(i)}(k+1)B(i) + R(i)\right)^{-1}B'(i)\bar{P}^{(i)}(k+1)A(i). \end{cases} \quad (4.1.15)$$

Then the infinite-time stochastic LQ problems are solvable with $u^*(k) = K(i)\,x(k)$, *and the optimal cost functions incurred by playing strategies* $u^*(k)$ *are* $x_0'P^{(i)}(0)x_0, i \in \Xi$.

4.2 Stochastic Nash Differential Games with Two Person

4.2.1 Finite-Time Horizon

4.2.1.1 Problem Formulation

On a probabilistic space $(\Omega, \mathcal{F}, \{\mathcal{F}_t\}_{t\geq 0}, \mathcal{P})$,, we consider a discrete-time markov jump linear systems of the following type:

$$\begin{cases} x(k+1) = [A(r_k)x(k) + B_1(r_k)u(k) + B_2(r_k)v(k)] + A_1(r_k)x(k)w(k), \\ x(0) = x_0 \in \mathbb{R}^n, \end{cases} \tag{4.2.1}$$

where $x(k) \in \mathbb{R}^n$ is the state, $u(k)$ and $v(k)$ represent the system control inputs, x_0 is a deterministic vector. r_k Denotes a time-varying markov chain taking values in $\Xi = \{1, \ldots, l\}$ with transition probability matrix $\rho(k) = [p_{ij}(k)], p_{ij}(k) = \mathrm{P}(r_{k+1} = j | r_k = i)$. The coefficients $A(r_k)$,$B_1(r_k)$,$A_1(r_k)$ are assumed to be constant matrices with appropriate dimensions. $w(k)$ is a one-dimensional standard Brownian motion. Assumpting that $w(k)$ is uncorrelated with $u(k)$ and $v(k)$, and is independent of the Markov chain r_k for $k \in N_T$. The initial value $r(0) = r_0$ is independent of the noise $w(k)$.

Let the optimal strategies for system (4.2.1) be given as

$$u(k) = K_1(r_k)x(k), v(k) = K_2(r_k)x(k) \tag{4.2.2}$$

The purpose of the two person Nash differential games is to find the feedback controls with constant matrix gain $K_\tau(r_k),\ \tau = 1, 2$ satisfying the following criteria

$$\begin{aligned} &J_\tau(u, v; x_0, i) \\ &= x'(T)M_\tau x(T) + \sum_{k=0}^{T-1} [x'(k)Q_\tau(r_k)x(k) + u'(k)R_{\tau 1}(r_k)u(k) + v'(k)R_{\tau 2}(r_k)v(k)], \tau = 1, 2 \end{aligned} \tag{4.2.3}$$

which also minimizing

$$J_1(u^*, v^*; x_0, i) \leq J_1(u, v^*; x_0, i), J_2(u^*, v^*; x_0, i) \leq J_2(u^*, v; x_0, i), \tag{4.2.4}$$

where all the weighting matrices $R_{\tau 1}(r_k) \in \mathcal{S}_l^m, R_{\tau 2}(r_k) \in \mathcal{S}_l^m, Q_\tau(r_k) \geq 0 \in \mathcal{S}_l^n$, $\tau = 1, 2$. In (4.2.1)– (4.2.3), when $r_k = i, i = 1, \ldots, l, A(r_k) = A(i)$, etc.

4.2.1.2 Main Result

With the help of stochastic LQ problems, it is not difficult to get the following Theorem 4.2.1:

Theorem 4.2.1 *The two person stochastic LQ differential games (4.2.1)–(4.2.3) in finite-time horizon are solvable with* $(u^*(\cdot), v^*(\cdot))$, *if and only if the following coupled Riccati equations*

$$\begin{cases} [A(i)+B_2(i)K_2(i)]'\bar{P}_1^{(i)}(k+1)[A(i)+B_2(i)K_2(i)]+Q_1(i)+A_1'(i)\bar{P}_1^{(i)}(k+1)A_1(i) \\ -P_1^{(i)}(k)+K_2'(i)R_{12}(i)K_2(i)+K_1'(i)B_1'(i)\bar{P}_1^{(i)}(k+1)(A(i)+B_2(i)K_2(i))=0, \\ P_1(T)=M_1, \\ B_1'(i)\bar{P}_1^{(i)}(k+1)B_1(i)+R_{11}(i)>0, \\ K_1(i)=-\left[B_1'(i)\bar{P}_1^{(i)}(k+1)B_1(i)+R_{11}(i)\right]^{-1}B_1'(i)\bar{P}_1^{(i)}(k+1)(A(i)+B_2(i)K_2(i)), \end{cases} \tag{4.2.4}$$

$$\begin{cases} [A(i)+B_1(i)K_1(i)]'\overline{P}_2^{(i)}(k+1)[A(i)+B_1(i)K_1(i)]+Q_2(i)+A_1'(i)\overline{P}_2^{(i)}(k+1)A_1(i) \\ -P_2^{(i)}(k)+K_1'(i)R_{21}(i)K_1(i)+K_2'(i)B_2'(i)\overline{P}_2^{(i)}(k+1)(A(i)+B_1(i)K_1(i))=0, \\ P_2(T)=M_2, \\ B_2'(i)\overline{P}_2^{(i)}(k+1)B_2(i)+R_{22}(i)>0, \\ K_2(i)=-\left[B_2'(i)\overline{P}_2^{(i)}(k+1)B_2(i)+R_{22}(i)\right]^{-1}B_2'(i)\overline{P}_2^{(i)}(k+1)(A(i)+B_1(i)K_1(i)), \end{cases} \tag{4.2.5}$$

admit a group of solutions $P(\cdot)=(P_1(\cdot),P_2(\cdot))\in\mathcal{S}_l^n\times\mathcal{S}_l^n$,*where* $P_1(\cdot)=(P_1^{(1)}(\cdot),\ldots,P_1^{(l)}(\cdot))$, $P_2(\cdot)=(P_2^{(1)}(\cdot),\ldots,P_2^{(l)}(\cdot))$, $(i,j\in\Xi)$,$\{P_\tau^{(i)}(k)\in\mathcal{S}_n\}$ *represents symmetric matrix indexed by the time k and the mode of operation i, and*

$$\overline{P}_\tau^{(i)}(k+1)=E(P_\tau^{r_{k+1}}(k+1))=\sum_{j=1}^{l}\pi_{ij}P_\tau^{(i)}(k+1)$$

Meanwhile, the explicit forms of the optimal strategies are

$$u^*(k)=\sum_{i=1}^{l}K_1(i)\chi_{r_k=i}(k)x(k),\ v^*(k)=\sum_{i=1}^{l}K_2(i)\chi_{r_k=i}(k)x(k),$$

and the optimal cost functions incurred by playing strategies are $J_\tau(u^*,v^*;x_0,i)=x_0'P_\tau^{(i)}(0)x_0\ ,\ \tau=1,2.$

Proof If (4.2.4) and (4.2.5) admit a group of solutions $P(\cdot) = (P_1(\cdot), P_2(\cdot)) \in \mathcal{S}_l^n \times \mathcal{S}_l^n$, let $v^*(k) = \sum_{i=1}^{l} K_2(i)\chi_{r_k=i}(k)x(k)$, and by substituting $v^*(k)$ into (4.2.1), we can obtain the following optimal control problem

$$
\begin{aligned}
&J_1(u, v^*; x_0, i) \\
&\quad = \mathbf{E}\left\{ x'(T)M_1x(T) + \sum_{k=0}^{T-1} \begin{bmatrix} x'(k)\left(Q_1(r_k) + K_2'(r_k)R_{12}(r_k)K_2(r_k)\right)x(k) \\ + u'(k)R_{11}(r_k)u(k) \end{bmatrix} |r_0 = i \right\},
\end{aligned}
$$

$s.t$

$$
\begin{cases} x(k+1) = [(A(r_k) + B_2(r_k)K_2(r_k))x(k) + B_1(r_k)u(k)] + A_1(r_k)x(k)w(k), \\ x(0) = x_0. \end{cases} \tag{4.2.6}
$$

The above optimal control problem is a standard stochastic LQ problem, so taking

$$
\begin{aligned}
&A(r_k) + B_2(r_k)K_2(r_k) \Rightarrow A(r_k), B_1(r_k) \Rightarrow B(r_k), A_1(r_k) \Rightarrow A_1(r_k), \\
&Q(r_k) + K_2'(r_k)R_{12}(r_k)K_2(r_k) \Rightarrow Q_1(r_k), R_{11}(r_k) \Rightarrow R(r_k).
\end{aligned}
$$

According to Theorem 4.1.1, we have

$$
u^*(k) = \sum_{i=1}^{l} K_1(i)\chi_{r_k=i}(k)x(k), i = 1, \cdots, l. \tag{4.2.7}
$$

Similarly, by substituting $u^*(k)$ into (4.2.1) we can obtain $v^*(k) = \sum_{i=1}^{l} K_2(i)\chi_{r_k=i}(k)x(k)$ is the optimal control strategy.

This completes the proof. □

4.2.2 *Infinite-Time Horizon*

4.2.2.1 Problem Formulation

We still consider the discrete-time markov jump linear systems (4.2.1). For convenience of description, we copy it as follows:

$$\begin{cases} x(k+1) = [A(r_k)x(k) + B_1(r_k)u(k) + B_2(r_k)v(k)] + A_1(r_k)x(k)w(k), \\ x(0) = x_0 \in \mathbb{R}^n. \end{cases} \tag{4.2.8}$$

Let the optimal strategies for system (4.2.8) be given as

$$u(k) = K_1(r_k)x(k), v(k) = K_2(r_k)x(k). \tag{4.2.9}$$

The purpose of the two person Nash differential games is to find the feedback controls $u(k) = K_1(r_k)x(k)$ and $v(k) = K_2(r_k)x(k)$ with constant matrix gain $K_\tau(r_k),\ \tau = 1,2$ satisfying the following criteria:

$$\begin{aligned} &J_\tau(u, v; x_0, i) \\ &= \sum_{k=0}^{\infty} [x'(k)Q_\tau(r_k)x(k) + u'(k)R_{\tau 1}(r_k)u(k) + v'(k)R_{\tau 2}(r_k)v(k)], \tau = 1,2, \end{aligned} \tag{4.2.10}$$

which also minimizing

$$J_1(u^*, v^*; x_0, i) \le J_1(u, v^*; x_0, i), J_2(u^*, v^*; x_0, i) \le J_2(u^*, v; x_0, i), \tag{4.2.11}$$

where all the weighting matrices $R_{\tau 1}(r_k) \in \mathcal{S}_l^m, R_{\tau 2}(r_k) \in \mathcal{S}_l^m$, $Q_\tau(r_k) \ge 0 \in \mathcal{S}_l^n, \tau = 1,2$. In (4.2.8) and (4.2.10), when $r_k = i, i = 1,\ldots,l, A(r_k) = A(i)$, etc.

4.2.2.2 Main Result

Assumption 4.2.1 *System (4.2.8) is mean-square stable.*

By the method used in the Nash stochastic differential games in finite-time horizon above, we can easily obtain the necessary and sufficient conditions as Theorem 4.2.3 for the equilibrium solution of the Nash stochastic differential games in infinite-time horizon, the proof is similar as Theorems 4.2.1 and 4.2.2, so it is not repeated herein.

Theorem 4.2.3 *Under the assumption 4.2.1, the two-person Nash stochastic differential games (4.2.8)–(4.2.11) in infinite-time horizon are solvable with* $(u^*(\cdot), v^*(\cdot))$, *if and only if the following coupled Riccati equations admit a group of solutions* $P(\cdot) = (P_1(\cdot), P_2(\cdot)) \in \mathcal{S}_l^n \times \mathcal{S}_l^n$, *where* $P_1(\cdot) = (P_1^{(1)}(\cdot),\ldots,P_1^{(l)}(\cdot))$, $P_2(\cdot) = (P_2^{(1)}(\cdot),\ldots,P_2^{(l)}(\cdot))$, $(i,j \in \Xi)$

$$\begin{cases} [A(i)+B_2(i)K_2(i)]'\overline{P}_1^{(i)}(k+1)[A(i)+B_2(i)K_2(i)]+Q_1'(i)Q_1(i) \\ +A_1'(i)\overline{P}_1^{(i)}(k+1)A_1(i)+K_2'(i)R_{12}(i)K_2(i)-P_1^{(i)}(k) \\ +K_1'(i)B_1'(i)\bar{P}_1^{(i)}(k+1)(A(i)+B_2(i)K_2(i))=0, \\ B_1'(i)\bar{P}_1^{(i)}(k+1)B_1(i)+R_{11}(i)>0, \\ K_1(i)=-\left[B_1'(i)\bar{P}_1^{(i)}(k+1)B_1(i)+R_{11}(i)\right]^{-1}B_1'(i)\bar{P}_1^{(i)}(k+1)(A(i)+B_2(i)K_2(i)), \end{cases} \tag{4.2.12}$$

$$\begin{cases} [A(i)+B_1(i)K_1(i)]'\overline{P}_2^{(i)}(k+1)[A(i)+B_1(i)K_1(i)]+Q_2'(i)Q_2(i) \\ +A_1'(i)\overline{P}_2^{(i)}(k+1)A_1(i)+K_1'(i)R_{21}(i)K_1(i)-P_2^{(i)}(k) \\ +K_2'(i)B_2'(i)\overline{P}_2^{(i)}(k+1)(A(i)+B_1(i)K_1(i))=0, \\ B_2'(i)\overline{P}_2^{(i)}(k+1)B_2(i)+R_{22}(i)>0, \\ K_2(i)=-\left[B_2'(i)\overline{P}_2^{(i)}(k+1)B_2(i)+R_{22}(i)\right]^{-1}B_2'(i)\overline{P}_2^{(i)}(k+1)(A(i)+B_1(i)K_1(i)). \end{cases} \tag{4.2.13}$$

Meanwhile, the explicit forms of the optimal strategies are

$$u^*(k)=\sum_{i=1}^{l}K_1(i)\chi_{r_k=i}(k)x(k),\ v^*(k)=\sum_{i=1}^{l}K_2(i)\chi_{r_k=i}(k)x(k).$$

and the optimal cost functions incurred by playing strategies are $J_\tau(u^*,v^*;x_0,i)=x_0'P_\tau^{(i)}(0)x_0$, $\tau=1,2$.

4.2.3 Two Person Zero-Sum Stochastic Differential Games

4.2.3.1 Finite Time Horizon

On a probabilistic space $(\Omega,\mathcal{F},\{\mathcal{F}_t\}_{t\geq 0},\mathcal{P})$, we consider a discrete-time markov jump linear systems of the following type:

$$\begin{cases} x(k+1)=A(r_k)x(k)+B_1(r_k)u_1(k)+B_2(r_k)u_2(k) \\ \qquad\qquad +[A_1(r_k)x(k)+C_1(r_k)u_1(k)+C_2(r_k)u_2(k)]w(k), \\ x(0)=x_0\in\mathbb{R}^n, \end{cases} \tag{4.2.14}$$

where $x(k)\in\mathbb{R}^n$ is the state, $u_1(k)$ and $u_2(k)$ represent the system control inputs, x_0 is a deterministic vector. r_k Denotes a time-varying markov chain taking values in $\Xi=\{1,\cdots,l\}$ with transition probability matrix $\rho(k)=[p_{ij}(k)],p_{ij}(k)=\mathrm{P}(r_{k+1}=j|r_k=i)$. The coefficients $A(r_k),B_1(r_k),B_2(r_k),C_1(r_k),C_2(r_k),A_1(r_k)$ are assumed to be constant matrices with appropriate dimensions. $w(k)$ is a one-dimensional standard Brownian motion. $w(k)$ is uncorrelated to $u_1(k)$ and $u_2(k)$, and is independent of the Markov chain r_k for $k \in N_T$. The initial value $r(0) = r_0$ is independent of the noise $w(k)$.

Let the optimal strategies for system (4.2.14) be given as

$$u_1(k) = K_1(r_k)x(k), u_2(k) = K_2(r_k)x(k).$$

The purpose of the two person zero-sum differential games is to find the feedback controls $u_1(k) = K_1(r_k)x(k)$ and $u_2(k) = K_2(r_k)x(k)$ with constant matrix gain $K_\tau(r_k),\ \tau = 1, 2$ satisfying the following criterion:

$$J(u_1, u_2; x_0, i) = \mathbf{E}\left\{\left[x'(T)Mx(T) + \sum_{k=0}^{T-1}\begin{pmatrix} x'(k)Q(r_k)x(k) + u_1'(k)R_1(r_k)u_1(k) \\ + u_2'(k)R_2(r_k)u_2(k)\end{pmatrix}\right]|r_0 = i\right\}, \tag{4.2.15}$$

which also minimizing

$$J\left(u_1^*, u_2; x_0, i\right) \le J\left(u_1^*, u_2^*; x_0, i\right) \le J\left(u_1, u_2^*; x_0, i\right),$$

where all the weighting matrices $R_1(r_k) \in \mathcal{S}_l^m, R_2(r_k) \in \mathcal{S}_l^m$, $Q(r_k) \ge 0 \in \mathcal{S}_l^n$.

In (4.2.14) and (4.2.15), when $r_k = i,\ i = 1, \ldots, l, A(r_k) = A(i)$, etc.

Theorem 4.2.4 *The two-person zero-sum stochastic differential games (4.2.14)–(4.2.15) in finite-time horizon are solvable with* $(u_1{}^*(\cdot), u_2{}^*(\cdot))$, *if and only if the following coupled Riccati equations admit a group of solutions* $P(\cdot) = (P^{(1)}(\cdot), \ldots, P^{(l)}(\cdot)) \in \mathcal{S}_l^n$, $(i \in \Xi)$

$$\begin{cases} \mathbf{N}(i,P) - P^{(i)}(k) - \mathbf{L}'(i,P)\mathbf{R}^{-1}(i,P)\mathbf{L}(i,P) = 0, \\ P(T) = M, \end{cases} \tag{4.2.16}$$

where,

$$\begin{cases} \mathbf{N}(i,P) = A'(i)\overline{P}^{(i)}(k+1)A(i) + Q(i) + A_1'(i)\overline{P}^{(i)}(k+1)A_1(i), \\ \mathbf{R}(i,P) = \begin{bmatrix} \Delta_{11}(i,P) + R_1(i) & \Delta_{12}(i,P) \\ \Delta_{21}(i,P) & \Delta_{22}(i,P) + R_2(i) \end{bmatrix}, \\ \Delta_{11}(i,P) + R_1(i) > 0, \Delta_{22}(i,P) + R_2(i) < 0 \\ \Delta_{mn}(i,P) = B_m'(i)\overline{P}^{(i)}(k+1)B_n(i) + C_m'(i)\overline{P}^{(i)}(k+1)C_n(i)(m,n = 1,2), \\ \mathbf{L}(i,P) = \begin{bmatrix} B_1'(i)\overline{P}^{(i)}(k+1)A(i) + C_1'(i)\overline{P}^{(i)}(k+1)A_1(i) \\ B_2'(i)\overline{P}^{(i)}(k+1)A(i) + C_2'(i)\overline{P}^{(i)}(k+1)A_1(i) \end{bmatrix}, \end{cases} \tag{4.2.17}$$

Meanwhile, the explicit forms of the optimal strategies are

$$\bar{u}^*(k) = (\,u_1^{*\prime}(k) \quad u_2^{*\prime}(k)\,)' = K(i)x(k) = -\mathbf{R}^{-1}(i,P)\mathbf{L}(i,P)x(k),$$

and the optimal cost functions incurred by playing strategies are $x_0'P^{(i)}(0)x_0$.

Proof The two-person zero-sum stochastic differential game is the special case of Nash stochastic differential game, so the proof of Theorem 4.2.4 can be referred to the Theorem 4.2.3, and it is not repeated herein.

4.2.3.2 Infinite-Time Horizon

We still consider the discrete-time markov jump linear systems (4.2.14), in order to discuss, we copy (4.2.14) as follows:

$$\begin{cases} x(k+1) = A(r_k)x(k) + B_1(r_k)u_1(k) + B_2(r_k)u_2(k) \\ \qquad\qquad + [A_1(r_k)x(k) + C_1(r_k)u_1(k) + C_2(r_k)u_2(k)]w(k), \\ x(0) = x_0 \in \mathbb{R}^n, \end{cases} \tag{4.2.18}$$

The purpose of the two person zero-sum differential games is to find the feedback controls $u_1(k) = K_1(r_k)x(k)$ and $u_2(k) = K_2(r_k)x(k)$ with constant matrix gain $K(r_k)$ satisfying the following criterion:

$$J(u_1,u_2;x_0,i) = \mathbf{E}\left\{\left[\sum_{k=0}^{\infty}\begin{pmatrix} x'(k)Q(r_k)x(k) + u_1'(k)R_1(r_k)u_1(k) \\ + u_2'(k)R_2(r_k)u_2(k) \end{pmatrix}\right]|r_0 = i\right\}, \tag{4.2.19}$$

which also minimizing

$$J(u_1^*,u_2;x_0,i) \le J(u_1^*,u_2^*;x_0,i) \le J(u_1,u_2^*;x_0,i),$$

where all the weighting matrices $R_1(r_k) \in \mathcal{S}_l^m, R_2(r_k) \in \mathcal{S}_l^m,\ Q(r_k) \ge 0 \in \mathcal{S}_l^n$. In (4.2.18) and (4.2.19), when $r_k = i, i = 1,\ldots,l, A(r_k) = A(i)$, etc.

Assumption 4.2.2 *Systems (4.2.18) is mean-square stable.*

By the method used in the zero-sum stochastic differential games in finite-time horizon above, we can easily obtain Theorem 4.2.5.

Theorem 4.2.4 *Under the assumption 4.2.2, the two-person zero-sum stochastic differential games (4.2.18)–(4.2.19) are solvable with* $(u_1{}^*(\cdot), u_2{}^*(\cdot))$, *if and only if* *the following coupled Riccati equations admit a group of solutions* $P(\cdot) = (P^{(1)}(\cdot), \cdots, P^{(l)}(\cdot)) \in \mathcal{S}_l^n, (i \in \Xi)$

$$\begin{cases} \mathbf{N}(i,P) - P^{(i)}(k) - \mathbf{L}'(i,P)\mathbf{R}^{-1}(i,P)\mathbf{L}(i,P) = 0, \\ \mathbf{N}(i,P) = A'(i)\overline{P}^{(i)}(k+1)A(i) + Q(i) + A_1'(i)\overline{P}^{(i)}(k+1)A_1(i), \\ \mathbf{R}(i,P) = \begin{bmatrix} \Delta_{11}(i,P) + R_1(i) & \Delta_{12}(i,P) \\ \Delta_{21}(i,P) & \Delta_{22}(i,P) + R_2(i) \end{bmatrix}, \\ \Delta_{mn}(i,P) = B_m'(i)\overline{P}^{(i)}(k+1)B_n(i) + C_m'(i)\overline{P}^{(i)}(k+1)C_n(i)(m,n=1,2), \\ \mathbf{L}(i,P) = \begin{bmatrix} B_1'(i)\overline{P}^{(i)}(k+1)A(i) + C_1'(i)\overline{P}^{(i)}(k+1)A_1(i) \\ B_2'(i)\overline{P}^{(i)}(k+1)A(i) + C_2'(i)\overline{P}^{(i)}(k+1)A_1(i) \end{bmatrix}. \end{cases} \tag{4.2.21}$$

Meanwhile, the explicit forms of the optimal strategies are

$$\bar{u}^*(k) = (\, u_1^{*\prime}(k) \quad u_2^{*\prime}(k)\,)' = K(i)x(k) = -\mathbf{R}^{-1}(i,P)\mathbf{L}(i,P)x(k),$$

and the optimal cost functions incurred by playing strategies are $x_0'P^{(i)}(0)x_0$.

4.3 Stackelberg Differential Games with Two Person

4.3.1 Finite-Time Horizon

4.3.1.1 Problem Formulation

We consider a discrete-time markov jump linear systems of the following type:

$$\begin{cases} x(k+1) = [A(r_k)x(k) + B(r_k)u(k) + C(r_k)v(k)] + A_1(r_k)x(k)w(k) \\ x(0) = x_0 \in \mathbb{R}^n \end{cases} \tag{4.3.1}$$

where $x(k) \in \mathbb{R}^n$ is the state, $u(k)$ and $v(k)$ represent the system control inputs, x_0 is a deterministic vector. r_k Denotes a time-varying markov chain taking values in $\Xi = \{1, \cdots, l\}$ with transition probability matrix $\rho(k) = [p_{ij}(k)], p_{ij}(k) = \mathrm{P}(r_{k+1} = j | r_k = i)$. The coefficients $A(r_k), B(r_k), C(r_k), A_1(r_k)$ are assumed to be constant matrices with appropriate dimensions. $w(k)$ is a one-dimensional standard Brownian motion. $w(k)$ is uncorrelated to $u(k)$ and $v(k)$, and is independent of the Markov chain r_k for $k \in N_T$ .The initial value $r(0) = r_0$ is independent of the noise $w(k)$.

Let the optimal strategies for system (4.3.1) be given as

$$u(k) = K_1(r_k)x(k), v(k) = K_2(r_k)x(k)$$

The performance functions are as follows:

$$\begin{aligned} &J_\tau(u, v; x_0, i) \\ &= \mathbf{E}\left\{x'(T)F_\tau x(T) + \sum_{k=0}^{T-1}[x'(k)Q_\tau(r_k)x(k) + u'(k)R_{\tau 1}(r_k)u(k) + v'(k)R_{\tau 2}(r_k)v(k)] | r_0 = i\right\}, \\ &\tau = 1, 2, \end{aligned} \tag{4.3.2}$$

where all the weighting matrices $R_{\tau 1}(r_k) \in \mathcal{S}_l^m, R_{\tau 2}(r_k) \in \mathcal{S}_l^m,\ Q_\tau(r_k) \geq 0 \in \mathcal{S}_l^n$, $\tau = 1, 2$.

In (4.3.1) and (4.3.2), when $r_k = i, i = 1, \cdots, l, A(r_k) = A(i)$, etc.

The definition of stackelberg equilibrium solution in finite-time horizon is as follows:

Definition 4.3.1 ([2]) For control strategy $u \in \mathcal{U}$, the optimal reaction set of the follower P_2 is:

$$\Re_2(u) = \{v^0 \in \mathcal{V} : J_2(u, v^0; x_0, i) \leq J_2(u, v; x_0, i)\}, \forall v \in \mathcal{V},$$

u^* is called the Stackelberg strategy of the leader P_1, if the following conditions hold:

$$\min_{v \in \Re_2(u^*)} J_1(u^*, v; x_0, i) \leq \min_{v \in \Re_2(u)} J_1(u, v; x_0, i), \forall u \in \mathcal{U}$$

4.3.1.2 Main Result

Stackelberg strategies of the discrete-time markov jump linear system are as the following theorem:

Theorem 4.3.1 *For system (4.3.1), if the following Riccati equations*:

$$\begin{cases} P_1^{(i)}(k) = H_1(P, i), P_1(T) = F_1, \\ P_2^{(i)}(k) = H_2(P, i), P_2(T) = F_2, \\ K_1(i) = -M^{-1}(P, i)L(P, i), \\ K_2(i) = \psi(P, i)(A(i) + B(i)K_1(i)), \\ M(P, i) > 0, C'(i)\overline{P}_2^{(i)}(k+1)C(i) + R_{22}(i) > 0, \end{cases} \tag{4.3.3}$$

with

$$\begin{aligned}
H_1(P,i) &= A'(i)\bar{P}_1^{(i)}(k+1)A(i) + A_1'(i)\bar{P}_1^{(i)}(k+1)A_1(i) + Q_1(i) + K_1'(i)L(P,i) \\
&\quad + A'(i)\psi'(P,i)\Big(C'(i)\overline{P}_1^{(i)}(k+1)C(i) + R_{12}(i)\Big)\psi(P,i)A(i) \\
&\quad + 2A'(i)\psi'(P,i)C'(i)\overline{P}_1^{(i)}(k+1)A(i), \\
H_2(P,i) &= A'(i)\overline{P}_2^{(i)}(k+1)A(i) + A_1'(i)\overline{P}_2^{(i)}(k+1)A_1(i) + Q_2(i) \\
&\quad + 2K_1'(i)B'(i)\overline{P}_2^{(i)}(k+1)A(i) + K_2'(i)C'(i)\overline{P}_2^{(i)}(k+1)(A(i)x(k) + B(i)K_1(i)) \\
&\quad + K_1'(i)\Big(B'(i)\overline{P}_2^{(i)}(k+1)B(i) + R_{21}(i)\Big)K_1(i), \\
\psi(P,i) &= -\Big[C'(i)\overline{P}_2^{(i)}(k+1)C(i) + R_{22}(i)\Big]^{-1}C'(i)\overline{P}_2^{(i)}(k+1), \\
M(P,i) &= B'(i)\overline{P}_1^{(i)}(k+1)B(i) + R_{11}(i) + 2B'(i)\overline{P}_1^{(i)}(k+1)C(i)\psi(P,i)B(i) \\
&\quad + B'(i)\psi'(P,i)\Big(C'(i)\overline{P}_1^{(i)}(k+1)C(i) + R_{11}(i)\Big)\psi(P,i)B(i), \\
L(P,i) &= B'(i)\overline{P}_1^{(i)}(k+1)A(i) + B'(i)\psi'(P,i)C'(i)\overline{P}_1^{(i)}(k+1)A(i) \\
&\quad + B'(i)\psi'(P,i)\Big(C'(i)\overline{P}_1^{(i)}(k+1)C(i) + R_{12}(i)\Big)\psi(P,i)A(i) \\
&\quad + B'(i)\overline{P}_1^{(i)}(k+1)C(i)\psi(P,i)A(i)
\end{aligned}$$

exist the solutions of $P(\cdot) = (P_1(\cdot), P_2(\cdot)) \in \mathcal{S}_l^n \times \mathcal{S}_l^n$, *where* $P_1(\cdot) = (P_1^{(1)}(\cdot), \ldots, P_1^{(l)}(\cdot))$, $P_2(\cdot) = (P_2^{(1)}(\cdot), \ldots, P_2^{(l)}(\cdot))$, *then the solutions of the Stackelberg game are:*

$$u^*(k) = \sum_{i=1}^{l} K_1(i)\chi_{r_k=i}(k)x(k),\ v^*(k) = \sum_{i=1}^{l} K_2(i)\chi_{r_k=i}(k)x(k).$$

Proof Given the strategy u implemented by the leader P_2, considering the optimization problem of the follower P_2. Let the value function $V_2(k,x) = x'(k)P_2^{(r_k)}(k)x(k)$, using the Itô lemma, we have:

$$\begin{aligned}
&\mathbf{E}[\Delta Y_2(k,x)|r_0 = i] \\
&= \mathbf{E}\Big[\Big(x'(k+1)P_2^{(r_{k+1})}(k)x(k+1) - x'(k)P_2^{(r_k)}(k)x(k)\Big)|r_0 = i\Big] \\
&= \mathbf{E}\Big[x'(k)\Big(A'(i)\bar{P}_2^{(i)}(k+1)A(i) + A_1'(i)\bar{P}_2^{(i)}(k+1)A_1(i) - P_2^{(i)}(k)\Big)x(k) \\
&\quad + 2u'(k)B'(i)\bar{P}_2^{(i)}(k+1)A(i)x(k) + 2v'(k)C'(i)\bar{P}_2^{(i)}(k+1)A(i)x(k) \\
&\quad + 2u'(k)B'(i)\bar{P}_2^{(i)}(k+1)C(i)v(k) + u'(k)B'(i)\bar{P}_2^{(i)}(k+1)B(i)u(k) \\
&\quad + v'(k)C(i)\bar{P}_2^{(i)}(k+1)C(i)v(k)\Big].
\end{aligned} \tag{4.3.4}$$

Based on the equations of $\sum_{k=0}^{T-1}[\Delta Y_2(k,x)] = \left(x(T)'P(T)x(T) - x(0)'\right.$ $\left.P^{(i)}(0)x(0)\right)$, we can get:

$$\begin{aligned}
&\mathbf{E}[\Delta Y_2(k,x)|r_0 = i] \\
&= \mathbf{E}\left\{\sum_{k=0}^{T-1}\left[x'(k)\left(A'(i)\bar{P}_2^{(i)}(k+1)A(i) + A_1'(i)\bar{P}_2^{(i)}(k+1)A_1(i) - P_2^{(i)}(k)\right)x(k)\right.\right. \\
&\quad + 2u'(k)B'(i)\bar{P}_2^{(i)}(k+1)A(i)x(k) + 2v'(k)C'(i)\bar{P}_2^{(i)}(k+1)A(i)x(k) \\
&\quad + 2u'(k)B'(i)\bar{P}_2^{(i)}(k+1)C(i)v(k) + u'(k)B'(i)\bar{P}_2^{(i)}(k+1)B(i)u(k) \\
&\quad \left.\left. + v'(k)C'(i)\bar{P}_2^{(i)}(k+1)C(i)v(k)\right]\right\} \\
&= \mathbf{E}[x'(T)P(T)x(T)] - x_0'P_2^{(i)}(0)x_0.
\end{aligned} \tag{4.3.5}$$

Substituting (4.3.5) into $J_2(u,\ v;\ x_0,\ i)$, we have

$$\begin{aligned}
J_2(u,v;x_0,i) &= -\mathbf{E}[x'(T)P_2(T)x(T)] + x_0'P_2^{(i)}(0)x_0 + \varepsilon[x'(T)Q_2(T)x(T)] \\
&\quad + \sum_{k=0}^{T-1}\left[x'(k)\left(A'(i)\bar{P}_2^{(i)}(k+1)A(i) + A_1'(i)\bar{P}_2^{(i)}(k+1)A_1(i) - P_2^{(i)}(k) + Q_1(i)\right)x(k)\right] \\
&\quad + \mathbf{E}\sum_{k=0}^{T-1}\left[2u'(k)B'(i)\bar{P}_2^{(i)}(k+1)A(i)x(k) + u'(k)\left(B'(i)\bar{P}_2^{(i)}(k+1)B(i) + R_{21}\right)u(k)\right. \\
&\quad + 2u'(k)B'(i)\bar{P}_2^{(i)}(k+1)C(i)v(k) + 2v'(k)C'(i)\bar{P}_2^{(i)}(k+1)A(i)x(k) \\
&\quad \left. + v'(k)C(i)\bar{P}_2^{(i)}(k+1)C(i)v(k) + v'(k)R_{22}(i)v(k)\right].
\end{aligned} \tag{4.3.6}$$

In (4.3.6), taking the derivative with respect to v, we have

$$\begin{aligned}
v(k) &= -\left[C'(i)\bar{P}_2^{(i)}(k+1)C(i) + R_{22}(i)\right]^{-1}C'(i)\bar{P}_2^{(i)}(k+1)(A(i)x(k) + B(i)u(k)) \\
&= \psi(i)(A(i)x(k) + B(i)u(k)),
\end{aligned} \tag{4.3.7}$$

where $\psi(i) = -\left[C'(i)\bar{P}_2^{(i)}(k+1)C(i) + R_{22}(i)\right]^{-1}C'(i)\bar{P}_2^{(i)}(k+1)$.

Now, consider the strategy of P_1, take the value function as $Y_1(k,$ $x) = x'(k)P_1^{(r_k)}(k)x(k)$, then

$$
\begin{aligned}
&\mathbf{E}[\Delta Y_1(k,x)|r_0=i]\\
&=\mathbf{E}\Bigg\{\sum_{k=0}^{T-1}\Big[x'(k)\Big(A'(i)\bar{P}_1^{(i)}(k+1)A(i)+A_1'(i)\bar{P}_1^{(i)}(k+1)A_1(i)-P_1^{(i)}(k)\Big)x(k)\\
&+2u'(k)B'(i)\bar{P}_1^{(i)}(k+1)A(i)x(k)+2v'(k)C'(i)\bar{P}_1^{(i)}(k+1)A(i)x(k)\\
&+2u'(k)B'(i)\bar{P}_1^{(i)}(k+1)C(i)v(k)+u'(k)B'(i)\bar{P}_1^{(i)}(k+1)B(i)u(k)\\
&+v'(k)C'(i)\bar{P}_1^{(i)}(k+1)C(i)v(k)\Big]\Bigg\}\\
&=\mathbf{E}[x'(T)P_1(T)x(T)]-x_0'P_1^{(i)}(0)x_0.
\end{aligned}
\tag{4.3.8}
$$

Substituting (4.3.8) into $J_1(u,\ v;\ x_0,\ i)$, we have

$$
\begin{aligned}
J_1(u,v;x_0,i)=&-\mathbf{E}[x'(T)P_1(T)x(T)]+x_0'P_1^{(i)}(0)x_0+\mathbf{E}[x'(T)Q_1(T)x(T)]\\
&+\sum_{k=0}^{T-1}\Big[x'(k)\Big(A'(i)\bar{P}_1^{(i)}(k+1)A(i)+A_1'(i)\bar{P}_1^{(i)}(k+1)A_1(i)-P_1^{(i)}(k)+Q_1(i)\Big)x(k)\Big]\\
&+\mathbf{E}\sum_{k=0}^{T-1}\Big[2u'(k)B'(i)\bar{P}_1^{(i)}(k+1)A(i)x(k)+u'(k)\Big(B'(i)\bar{P}_1^{(i)}(k+1)B(i)+R_{11}(i)\Big)u(k)\\
&+2u'(k)B'(i)\bar{P}_1^{(i)}(k+1)C(i)v(k)+2v'(k)C'(i)\bar{P}_1^{(i)}(k+1)A(i)x(k)\\
&+v'(k)C'(i)\bar{P}_1^{(i)}(k+1)C(i)v(k)+v'(k)R_{12}(i)v(k)\Big].
\end{aligned}
\tag{4.3.9}
$$

Substituting (4.3.7) into (4.3.9), we have

$$
\begin{aligned}
J_1(u,v^*;x_0,i)=&-\mathbf{E}[x'(T)P_1(T)x(T)]+x_0'P_1^{(i)}(0)x_0+\mathbf{E}[x'(T)Q_1(T)x(T)]\\
&+\sum_{k=0}^{T-1}\Big\{x'(k)\Big[A'(i)\overline{P}_1^{(i)}(k+1)A(i)+A_1'(i)\overline{P}_1^{(i)}(k+1)A_1(i)-P_1^{(i)}(k)+Q_1(i)\\
&+A'(i)\psi'(P,i)\Big(C'(i)\overline{P}_1^{(i)}(k+1)C(i)+R_{12}(i)\Big)\psi(P,i)A(i)\\
&+2A'(i)\psi'(P,i)C'(i)\overline{P}_1^{(i)}(k+1)A(i)\Big]x(k)\Big\}\\
&+\mathbf{E}\sum_{k=0}^{T-1}\Big\{2u'(k)[B'(i)\overline{P}_1^{(i)}(k+1)A(i)+2B'(i)\psi'(i)C'(i)\overline{P}_1^{(i)}(k+1)A(i)\\
&+2B'(i)\psi'(i)\Big(C'(i)\overline{P}_1^{(i)}(k+1)C(i)+R_{12}(i)\Big)\psi(i)A(i)\\
&+2B'(i)\overline{P}_1^{(i)}(k+1)C(i)\psi(i)A(i)]\\
&\times x(k)+u'(k)[B'(i)\overline{P}_1^{(i)}(k+1)B(i)+R_{11}(i)+2B'(i)\overline{P}_1^{(i)}(k+1)C(i)\psi(i)B(i)\\
&+B'(i)\psi'(i)\Big(C'(i)\overline{P}_1^{(i)}(k+1)C(i)+R_{11}(i)\Big)\psi(i)B(i)]u(k)\Big\}.
\end{aligned}
\tag{4.3.10}
$$

Using completion square method to (4.3.10), we get

$$J_1(u, v^*; x_0, i) = -\mathbf{E}[x'(T)P_1(T)x(T)] + x_0'P_1^{(i)}(0)x_0 + \mathbf{E}[x'(T)F_1x(T)] + \sum_{k=0}^{T-1}[x'(k)\left(H_1(i) - P_1^{(i)}(k)\right)x(k) + (u(k) - K_1(i)x(k))M(P,i)(u(k) - K_1(i)x(k))]. \tag{4.3.11}$$

According to (4.3.3), $M(P, i) > 0$, we have

$$J_1(u, v^*; x_0, i) \geq J_1(u^*, v^*; x_0, i) = x_0'P_1^{(i)}(0)x_0 \tag{4.3.12}$$

then, $u^*(k) = K_1(i)x(k)$.

Substituting (4.3.12) into (4.3.6), we have

$$\begin{aligned} J_2(u^*, v; x_0, i) = {} & -\mathbf{E}[x'(T)P_2(T)x(T)] + x_0'P_2^{(i)}(0)x_0 + \mathbf{E}[x'(T)F_2x(T)] \\ & + \sum_{k=0}^{T-1}[x'(k)\left(H_2(i) - P_2^{(i)}(k)\right)x(k) \\ & + (v(k) - K_2(i)x(k))\left(C'(i)\overline{P}_2^{(i)}(k+1)C(i) + R_{22}(i)\right) \\ & \times (v(k) - K_2(i)x(k))]. \end{aligned}$$

According to (4.3.3), in view of $C'(i)\overline{P}_2^{(i)}(k+1)C(i) + R_{22}(i) > 0$, we have

$$J_2(u^*, v; x_0, i) \geq J_2(u^*, v^*; x_0, i) = x_0'P_2^{(i)}(0)x_0.$$

Then, $v^*(k) = \psi(i)(A(i) + B(i)K_1(i))x(k) = K_2(i)x(k)$.

This completes the proof. □

4.3.2 *Infinite-Time Horizon*

We still consider the discrete-time markov jump linear systems (4.3.1), in order to discuss, we copy the formula (4.3.1) as follows:

$$\begin{cases} x(k+1) = [A(r_k)x(k) + B(r_k)u(k) + C(r_k)v(k)] + A_1(r_k)x(k)w(k), \\ x(0) = x_0 \in \mathbb{R}^n. \end{cases} \tag{4.3.13}$$

Let the optimal strategies for system (4.3.13) be given as

$$u(k) = K_1(r_k)x(k), v(k) = K_2(r_k)x(k).$$

The performance functions are as follows:

$$\begin{aligned} &J_\tau(u, v; x_0, i) \\ &= \mathbf{E}\left\{\sum_{k=0}^{T}[x'(t)Q_\tau(r_k)x(k) + u'(t)R_{\tau 1}(r_k)u(k) + v'(k)R_{\tau 2}(r_k)v(k)]|r_0 = i\right\}, \tau = 1, 2, \end{aligned} \tag{4.3.14}$$

where all the weighting matrices $R_{\tau 1}(r_k) \in \mathcal{S}_l^m, R_{\tau 2}(r_k) \in \mathcal{S}_l^m,\ \ Q_\tau(r_k) \geq 0 \in \mathcal{S}_l^n, \tau = 1, 2.$

In (4.3.13) and (4.3.14), when $r_k = i, i = 1, \cdots, l, A(r_k) = A(i), \text{etc.}$

Definition 4.3.2 ([2]) For control strategy $u \in \mathcal{U}$, the optimal reaction set of the follower P_2 is:

$$\Re_2(u) = \{v^0 \in \mathcal{V} : J_2(u, v^0; x_0, i) \leq J_2(u, v; x_0, i)\}, \forall v \in \mathcal{V},$$

u^* is called the Stackelberg of the leader P_1, if the following conditions hold:

$$\min_{v \in \Re_2(u^*)} J_1(u^*, v; x_0, i) \leq \min_{v \in \Re_2(u)} J_1(u, v; x_0, i), \forall u \in \mathcal{U}.$$

Assumption 4.3.1 *Systems (4.3.13) is mean-square stable.*

By the method used in the stackelberg stochastic differential games in finite-time horizon above, we can easily obtain the sufficient conditions as Theorem 4.3.2 for the equilibrium solution of the stackelberg stochastic differential games in infinite-time horizon.

Theorem 4.3.2 *Under the assumption 4.3.1, for system (4.3.13) and (4.3.14), if the following Riccati equations*

$$\begin{cases} P_1^{(i)}(k) = H_1(P, i), P_2^{(i)}(k) = H_2(P, i), \\ K_1(i) = -M^{-1}(P, i)L(P, i), \\ K_2(i) = \psi(P, i)(A(i) + B(i)K_1(i)), \\ M(P, i) > 0, C'(i)\overline{P}_2^{(i)}(k+1)C(i) + R_{22}(i) > 0, \end{cases} \tag{4.3.15}$$

with

$$\begin{aligned} H_1(P, i) = &A'(i)\overline{P}_1^{(i)}(k+1)A(i) + A_1'(i)\overline{P}_1^{(i)}(k+1)A_1(i) + Q_1(i) + K_1'(i)L(P, i) \\ &+ A'(i)\psi'(P, i)\Big(C'(i)\overline{P}_1^{(i)}(k+1)C(i) + R_{12}(i)\Big)\psi(P, i)A(i) \\ &+ 2A'(i)\psi'(P, i)C'(i)\overline{P}_1^{(i)}(k+1)A(i), \end{aligned}$$

$$H_2(P,i) = A'(i)\overline{P}_2^{(i)}(k+1)A(i) + A_1'(i)\overline{P}_2^{(i)}(k+1)A_1(i) + Q_2(i)$$
$$+ 2K_1'(i)B'(i)\overline{P}_2^{(i)}(k+1)A(i) + K_2'(i)C'(i)\overline{P}_2^{(i)}(k+1)(A(i)x(k) + B(i)K_1(i))$$
$$+ K_1'(i)\left(B'(i)\overline{P}_2^{(i)}(k+1)B(i) + R_{21}(i)\right)K_1(i),$$

$$\psi(P,i) = -\left[C'(i)\overline{P}_2^{(i)}(k+1)C(i) + R_{22}(i)\right]^{-1} C'(i)\overline{P}_2^{(i)}(k+1),$$
$$M(P,i) = B'(i)\overline{P}_1^{(i)}(k+1)B(i) + R_{11}(i) + 2B'(i)\overline{P}_1^{(i)}(k+1)C(i)\psi(P,i)B(i)$$
$$+ B'(i)\psi'(P,i)\left(C'(i)\overline{P}_1^{(i)}(k+1)C(i) + R_{11}(i)\right)\psi(P,i)B(i),$$
$$L(P,i) = B'(i)\overline{P}_1^{(i)}(k+1)A(i) + B'(i)\psi'(P,i)C'(i)\overline{P}_1^{(i)}(k+1)A(i)$$
$$+ B'(i)\psi'(P,i)\left(C'(i)\overline{P}_1^{(i)}(k+1)C(i) + R_{12}(i)\right)\psi(P,i)A(i)$$
$$+ B'(i)\overline{P}_1^{(i)}(k+1)C(i)\psi(P,i)A(i)$$

exist $P = (P_1, P_2) \in \mathcal{S}_l^n \times \mathcal{S}_l^n$, *where* $P_1 = (P_1^{(1)}, P_1^{(2)} \cdots, P_1^{(l)})$, $P_2 = (P_2^{(1)}, P_2^{(2)} \cdots, P_2^{(l)})$, *then the solutions of the Stackelberg game (4.3.13) and (4.3.14) are:*

$$u^*(k) = \sum_{i=1}^{l} K_1(i)\chi_{r_k=i}(k)x(k), v^*(k) = \sum_{i=1}^{l} K_2(i)\chi_{r_k=i}(k)x(k).$$

4.4 Summary

The Nash equilibrium, saddle point equilibrium, and stackelberg equilibrium strategy for the discrete time stochastic Markov jump linear systems in the finite-time horizon and infinite-time horizon are discussed respectively in this chapter. The optimal strategies and the optimal control values are obtained.

References

1. Wen-Xin, Cai, Yang-Wang, Fang, Rui, Li, & You-Li, Wu. (2012). Optimal control for Markov jump linear systems. *Systems Engineering and Electronics, 34*(7), 1458–1462.
2. Freilng, G., Jank, G., & Lee, S. R. (2001). Existence and uniqueness of open-loop Stackelberg equilibria in linear-quadratic differential games. *Journal of Optimization Theory and Applications, 110*(3), 515–544.

3. Willems, J. L., & Willems, J. C. (1976). Feedback stabilizability for stochastic systems with state and control dependent noise [J]. *Automatica*, *12*(3), 277–283.
4. Hou, T., Zhang, W., Ma, H. (2010). Finite horizon H_2/H_∞ control for discrete-time stochastic systems with markovian jumps and multiplicative noise [J]. *IEEE Transactions on Automatic Control*, *55*(5), 1185–1191.

Chapter 5
Stochastic Differential Game of Stochastic Markov Jump Singular Systems

Singular systems are a class of more generalized dynamic systems with a wide application background, which appeared in large number of practical system models, such as the power systems, economic systems, constrained robots, electronic networks and aerospace systems, so its research has important theoretical and practical value. This chapter attempts to extend the differential game theory study of Markov jump linear systems discussed in Chap. 3 to stochastic Markov jump singular systems, which covers the saddle point equilibrium theory of two person zero-sum games, the Nash equilibrium theory and Stackelberg game theory of two person nonzero-sum game, and the existence conditions, strategy design methods and algorithms of the equilibrium strategies are analyzed.

5.1 Stochastic LQ Problems—Differential Games of One Player

Stochastic linear quadratic control problems were abbreviated as stochastic LQ problem, which originate from the work of Wonham (1968) [1], and then had attracted great attention of many researchers (see [2–4] and references therein). Recently, stochastic LQ problem has been studied widespread; its theoretical basis has continuous improvement, and has been widely used in engineering, economics, management and other areas.

This section discusses the stochastic LQ problems for stochastic Markov jump singular system in finite-time horizon and infinite-time horizon, which laid the foundation for further study.

C.-k. Zhang et al., *Non-cooperative Stochastic Differential Game Theory of Generalized Markov Jump Linear Systems*, Studies in Systems, Decision and Control 67, DOI 10.1007/978-3-319-40587-2_5

5.1.1 Preliminaries

5.1.1.1 Stability of the Stochastic Markov Jump Singular Systems

First of all, let $(\Omega, \mathcal{F}, \{\mathcal{F}_t\}_{t\geq 0}, \mathcal{P})$ be a probability space with a natural filtration $\{\mathcal{F}_t\}_{t\geq 0}$ and $\mathbf{E}[\cdot]$ be the expectation operator with respect to the probability measure. Let $w(t)$ be a one-dimensional standard Brownian motion defined on the probability space and r_t be a right-continuous Markov chain independent of $w(t)$ and taking values in a finite set $\Xi = \{1, \ldots, l\}$ with transition probability matrix $\Pi = (\pi_{ij})_{l\times l}$ given by

$$\Pr\{r_{t+\Delta} = j | r_t = i\} = \begin{cases} \pi_{ij}\Delta + o(\Delta), & \text{if } i \neq j, \\ 1 + \pi_{ii}\Delta + o(\Delta), & \text{if } i = j, \end{cases} \tag{5.1.1}$$

where $\Delta > 0$, $\lim_{\Delta\to 0} o(\Delta)/\Delta = 0$, and $\pi_{ij} > 0$ $(i, j \in \Xi,\ i \neq j)$ denotes the transition rate from mode i at time t to mode j at time $t+h$, and $\pi_{ii} = -\sum_{j=1, j\neq i}^{l} \pi_{ij}$.

For a given Hilbert space $\mathcal{H}$ with the norm $\|\cdot\|_{\mathcal{H}}$, define the Banach space

$$L^2_{\mathcal{F}}(0, T; \mathcal{H}) = \left\{ \phi(\cdot) \left| \begin{array}{l} \phi(\cdot) \text{ is an } \mathcal{F}_t\text{-adapted, } \mathcal{H}\text{-valued measurable} \\ \text{process on } [0, T] \text{ and } \mathbf{E}\int_0^T \|\phi(t,\omega)\|^2_{\mathcal{H}} dt < +\infty, \end{array} \right. \right\},$$

with the norm

$$\|\phi(\cdot)\|_{\mathcal{F},2} = \left(\mathbf{E} \int_0^T \|\phi(t,\omega)\|^2_{\mathcal{H}} dt \right)^{\frac{1}{2}}.$$

Consider the following n-dimensional stochastic Markov jump singular systems

$$\begin{cases} Edx(t) = A(r_t)x(t)dt + F(r_t)x(t)dw(t), \\ x(0) = x_0, \end{cases} \tag{5.1.2}$$

where $x(t) \in \mathbb{R}^n$ is the system state, $(x_0, r_0) \in \mathbb{R}^n \times \Xi$ is the initial state, $A(r_t) = A(i)$, $F(r_t) = F(i)$, when $r_t = i$, $i \in \Xi$, are $n \times n$ matrices, $E \in \mathbb{R}^{n\times n}$ is a known singular matrix with $0 < rank(E) = k \leq n$. For simplicity, we also write $A_i = A(i)$, $F_i = F(i)$, etc. for all $i \in \Xi$ where there is no ambiguity.

In order to guarantee the existence and uniqueness of the solution to system (5.1.2), we give the following lemma [5].

Lemma 5.1.1 *For every $i \in \Xi$, if there are a pair of nonsingular matrices $M_i \in \mathbb{R}^{n\times n}$ and $N_i \in \mathbb{R}^{n\times n}$ for the triplet (E, A_i, F_i) such that one of the following conditions is satisfied, then* (5.1.2) *has a unique solution.*

(i) $M_iEN_i = \begin{bmatrix} I_k & 0 \\ 0 & 0 \end{bmatrix}$, $M_iA_iN_i = \begin{bmatrix} A_i & 0 \\ 0 & I_{n-k} \end{bmatrix}$, $M_iF_iN_i = \begin{bmatrix} F_i & B_i \\ 0 & C_i \end{bmatrix}$;

(ii) $M_iEN_i = \begin{bmatrix} I_k & 0 \\ 0 & 0 \end{bmatrix}$, $M_iA_iN_i = \begin{bmatrix} A_i & B_i \\ 0 & C_i \end{bmatrix}$, $M_iF_iN_i = \begin{bmatrix} F_i & 0 \\ 0 & I_{n-k} \end{bmatrix}$;

where $B_i \in \mathbb{R}^{k\times(n-k)}$ *and* $C_i \in \mathbb{R}^{(n-k)\times(n-k)}$.

Now, we are in a position to give the main result of the stability of stochastic Markov jump singular systems. Firstly, we recall the following definitions for system (5.1.2) [5].

Definition 5.1.1 Stochastic Markov jump singular system (5.1.2) is said to be

(i) regular if $\det(sE - A_i)$ is not identically zero for all $i \in \Xi$;
(ii) impulse free if $^\circ(\det(sE - A_i)) = rank(E)$ for all $i \in \Xi$;
(iii) mean square stable if for any initial condition $(x_0, r_0) \in \mathbb{R}^n \times \Xi$, we have $\lim_{t\to\infty} \mathbf{E}\|x(t)\|^2 = 0$;
(iv) mean square admissible if it is regular, impulse free and stable in mean square sense.

The following lemma presents the generalized Itô formula for Markov-modulated processes [6].

Lemma 5.1.2 *Given an n-dimensional process* $x(\cdot)$ *satisfying*

$$dx(t) = b(t, x(t), r_t)dt + \sigma(t, x(t), r_t)dw(t).$$

And a number of functions $\varphi(\cdot,\cdot,i) \in C^2([0,T]\times\mathbb{R}^n)$, $i = 1,\ldots,l$, *we have*

$$\mathbf{E}\{\varphi(T, x(T), r_T) - \varphi(s, x(s), r_s)|r_s = i\} = \mathbf{E}\left\{\int_s^T \Gamma\varphi(t, x(t), r_t)dt|r_s = i\right\},$$

where

$$\begin{aligned}\Gamma\varphi(t,x,i) &= \varphi_t(t,x,i) + b'(t,x,i)\varphi_x(t,x,i) \\ &\quad + \frac{1}{2}tr[\sigma'(t,x,i)\varphi_{xx}(t,x,i)\sigma(t,x,i)] + \sum_{j=1}^{l}\pi_{ij}\varphi(t,x,j).\end{aligned}$$

The following lemma generalized the results of stochastic singular systems presented in [7] to stochastic Markov jump singular systems.

Lemma 5.1.3 *Stochastic Markov jump singular system* (5.1.2) *is mean square admissible if there exist matrices* $P_i \in \mathbb{R}^{n\times n} > 0$, *such that the following coupled linear matrix inequalities (LMIs) hold for each* $i \in \Xi$

$$A_i'P_iE + E'P_iA_i + F_i'P_iF_i + \sum_{j=1}^{l}\pi_{ij}E'P_jE < 0. \quad (5.1.3)$$

Proof Under the conditions of the lemma, we see that the system (5.1.2) is of regularity and absence of impulses. Now we will show the mean square stability of the system (5.1.2). Consider a Lyapunov function candidate defined as follows:

$$V(t,x(t),r_t) = x'(t)E'P(r_t)Ex(t),$$

where $P(i)$ a symmetric matrix. □

Let $\mathcal{L}$ be be the infinitesimal generator. By Lemma 5.1.2, we have the stochastic differential as

$$dV(t,x(t),r_t) = \mathcal{L}V(x(t),r_t)dt + x'(t)[F(r_t)P(r_t)E + E'P(r_t)F(r_t)]x(t)dw(t), \tag{5.1.4}$$

where

$$\mathcal{L}V(t,x(t),i) = x'(t)[A_i'P_iE + E'P_iA_i + F_i'P_iF_i + \sum_{j=1}^{l} \pi_{ij}E'P_jE]x(t). \tag{5.1.5}$$

From (5.1.3) we have $\mathcal{L}V(t,x(t),i) < 0$. With the similar techniques in the work in [8], it can be seen that $\lim_{t\to\infty} \mathbf{E}\|x(t)\|^2 = 0$. Therefore, system (5.1.2) is mean square admissible. This completes the proof.

5.1.2 LQ Problem of Stochastic Markov Jump Singular Systems

We consider finite-time horizon and infinite-time horizon LQ problems of the stochastic Markov jump singular systems in the following section, respectively.

5.1.2.1 Finite-Time Horizon LQ Problem

Consider the following continuous-time time-varying stochastic Markov jump singular systems with state- and control-dependent noise

$$\begin{cases} Edx(t) = [A(t,r_t)x(t) + B(t,r_t)u(t)]dt + [C(t,r_t)x(t) + D(t,r_t)u(t)]dw(t), \\ x(0) = x_0, \end{cases} \tag{5.1.6}$$

where $(x_0, r_0) \in \mathbb{R}^n \times \Xi$ is the initial state, and an admissible control $u(\cdot)$ is an $\mathcal{F}_t$-adapted, $\mathbb{R}^m$-valued measurable process on $[0, T]$. The set of all admissible controls is denoted by $\mathcal{U}_{ad} = L^2_{\mathcal{F}}(0, T; \mathbb{R}^m)$. The solution $x(\cdot)$ of the Eq. (5.1.6) is called the response of the control $u(\cdot) \in \mathcal{U}_{ad}$, and $(x(\cdot), u(\cdot))$ is called an admissible pair. Here,$w(t)$ is a one-dimensional standard Brownian motion on $[0, T]$ (with $w(0) = 0$). Note that we assumed the Brownian motion to be one-dimensional just for simplicity. There is no essential difficulty in the analysis below for the multi-dimensional case.

For each (x_0, i) and $u(\cdot) \in \mathcal{U}_{ad}$, the associated cost is

$$J(x_0, i; u(\cdot)) = \mathbf{E}\left\{\int_0^T \begin{bmatrix} x(t) \\ u(t) \end{bmatrix}' \begin{bmatrix} Q(t, r_t) & L(t, r_t) \\ L'(t, r_t) & R(t, r_t) \end{bmatrix} \begin{bmatrix} x(t) \\ u(t) \end{bmatrix} dt + x'(T)H(r_T)x(T) | r_0 = i\right\}, \tag{5.1.7}$$

In (5.1.6) and (5.1.7), $A(t, r_t) = A_i(t)$, etc. whenever $r_t = i$, and $H(r_T) = H_i$ whenever $r_T = i$, whereas $A_i(\cdot)$ etc. are given matrix-valued functions and H_i are given matrices, $i = 1, \ldots, l$. The objective of the optimal control problem is to minimize the cost function $J(x_0, i; u(\cdot))$, for a given $(x_0, i) \in \mathbb{R}^n \times \Xi$, over all $u(\cdot) \in \mathcal{U}_{ad}$. The value function is defined as

$$V(x_0, i) = \inf_{u(\cdot) \in \mathcal{U}_{ad}} J(x_0, i; u(\cdot)). \tag{5.1.8}$$

Definition 5.1.2 The LQ problem (5.1.6)–(5.1.8) is called well-posed if

$$V(x_0, i) > -\infty, \quad \forall x_0 \in \mathbb{R}^n, \quad \forall i = 1, \ldots, l.$$

An admissible pair $(x^*(\cdot), u^*(\cdot))$ is called optimal (with respect to the initial condition (x_0, i)) if $u^*(\cdot)$ achieves the infimum of $J(x_0, i; u(\cdot))$.

The following basic assumption will be imposed in this section.

Assumption 5.1 The data appearing in the LQ problem (5.1.6)–(5.1.8) satisfy, for every i,

$$\begin{cases} A_i(\cdot),\ C_i(\cdot) \in L^\infty(0,\ T;\ \mathbb{R}^{n\times n}), \\ B_i(\cdot),\ D_i(\cdot) \in L^\infty(0,\ T;\ \mathbb{R}^{n\times m}), \\ Q_i(\cdot) \in L^\infty(0,\ T;\ \mathcal{S}^n), \\ L_i(\cdot) \in L^\infty(0,\ T;\ \mathbb{R}^{n\times m}), \\ R_i(\cdot) \in L^\infty(0,\ T;\ \mathcal{S}^m), \\ H_i \in \mathcal{S}^n. \end{cases}$$

Now we introduce a type of coupled differential Riccati equations associated with the LQ problem (5.1.6)–(5.1.8).

Definition 5.2 The following system of constrained differential equations

$$\begin{cases} E'\dot{P}_i(t)E + E'P_i(t)A_i(t) + A_i'(t)P_i(t)E + C_i'(t)P_i(t)C_i(t) + Q_i(t) + \sum_{j=1}^{l} \pi_{ij}E'P_j(t)E \\ -\left[E'P_i(t)B_i(t) + C_i'(t)P_i(t)D_i(t) + L_i(t)\right]\left[R_i(t) + D_i'(t)P_i(t)D_i(t)\right]^{-1} \\ \times\left[B_i'(t)P_i(t)E + D_i'(t)P_i(t)C_i(t) + L_i'(t)\right] = 0, \\ E'P_i(T)E = H_i, \\ R_i(t) + D_i'(t)P_i(t)D_i(t) > 0, \quad a.e.\ t \in [0,T], \quad i = 1,\ldots,l. \end{cases} \tag{5.1.9}$$

is called a system of coupled generalized differential Riccati equations (CGDREs).

Theorem 5.1.1 *If the CGDREs* (5.1.9) *admit a solution* $(P_1(\cdot),\ldots,P_l(\cdot)) \in C^1(0,T;\mathcal{S}_l^n)$, *then the finite-time horizon LQ problem* (5.1.6)–(5.1.8) *is well-posed. Moreover, the corresponding optimal feedback control law with respect to the initial* $(x_0,i) \in \mathbb{R}^n \times \Xi$ *is determined by the following:*

$$u^*(t) = \sum_{i=1}^{l} K_i(t)\chi_{r_t=i}(t)x(t), \tag{5.1.10}$$

where $K_i(t) = -\left[R_i(t) + D_i'(t)P_i(t)D_i(t)\right]^{-1}\left[B_i'(t)P_i(t)E + D'(t)_iP_i(t)C_i(t) + L_i'(t)\right]$. *Furthermore, the value function is uniquely determined by* $(P_1(\cdot),\ldots,P_l(\cdot)) \in C^1(0,T;\mathcal{S}_l^n)$:

$$V(x_0,i) = J(x_0,i;u^*(\cdot)) = x_0'E'P_i(0)Ex_0, \quad i = 1,\ldots,l. \tag{5.1.11}$$

Proof Given $P_i(\cdot) \in C^1(0,T;\mathcal{S}^n)$ and let $\varphi(t,x,i) = x'E_i'P_i(t)E_ix$. Applying the generalized Itô's formula to the linear system (5.1.6), we obtain

$$\begin{aligned} &\mathbf{E}[x'(T)H_{r_T}x(T)] - x_0'E'P_i(0)Ex_0 \\ &= \mathbf{E}[x'(T)E'P_{r_T}(T)Ex(T) - x'(0)E'P(r_0)Ex(0)|r_0 = i] \\ &= \mathbf{E}\{\varphi(T,x(T),r_T) - \varphi(0,x(0),r_0)|r_0 = i\} \\ &= \mathbf{E}\left\{\int_0^T \Gamma\varphi(t,x(t),r_t)dt|r_0 = i\right\}, \end{aligned}$$

where

$$\begin{aligned}\Gamma\varphi(t,x,i) &= \varphi_t(t,x,i) + b'(t,x,u,i)\varphi_x(t,x,i)\\ &\quad + \frac{1}{2}tr[\sigma'(t,x,u,i)\varphi_{xx}(t,x,i)\sigma(t,x,u,i)] + \sum_{j=1}^{l}\pi_{ij}V(t,x,j)\\ &= x'\left[E'\dot{P}_i(t)E + E'P_i(t)A_i(t) + A_i'(t)P_i(t)E + C_i'(t)P_i(t)C_i(t) + \sum_{j=1}^{l}\pi_{ij}E'P_j(t)E\right]x\\ &\quad + 2u'\left[B_i'(t)P_i(t)E + D_i'(t)P_i(t)C_i(t)\right]x + u'D_i'(t)P_i(t)D_i(t)u.\end{aligned}$$

Adding this to (5.1.7) and, provided $R_i(t) + D_i'(t)P_i(t)D_i(t) > 0$, using the square completion technique, we have

$$\begin{aligned}J(x_0,i;u(\cdot)) &= x_0'E'P_i(0)Ex_0 + \mathbf{E}\left\{\int_0^T [x(t)\big(E'\dot{P}_i(t)E + E'P_i(t)A_i(t) + A_i'(t)P_i(t)E\right.\\ &\quad + Q_i(t) + C_i'(t)P_i(t)C_i(t) + \sum_{j=1}^{l}\pi_{ij}E'P_j(t)E\Big)x(t)\\ &\quad + 2u'\big(B_i'(t)P_i(t)E_i + D_i'(t)P_i(t)C_i(t) + L_i'(t)\big)x(t)\\ &\quad \left. + u'(t)\big(R_i(t) + D_i'(t)P_i(t)D_i(t)\big)u(t)\big]dt|r_0 = i\right\}\\ &\quad + \mathbf{E}\{x'(T)(H_i - E'P_i(T)E)x(T)\}\\ &= x_0'E'P_i(0)Ex_0 + \mathbf{E}\left\{\int_0^T [x(t)\big(E'\dot{P}_i(t)E + E'P_i(t)A_i(t) + A_i'(t)P_i(t)E + Q_i(t)\right.\\ &\quad + C_i'(t)P_i(t)C_i(t) + \sum_{j=1}^{l}\pi_{ij}E'P_j(t)E - \big(EP_i(t)B_i(t) + C_i'(t)P_i(t)D_i(t) + L_i(t)\big)\\ &\quad \times \big(R_i(t) + D_i'(t)P_i(t)D_i(t)\big)^{-1}\big(B_i'(t)P_i(t)E_i + D_i'(t)P_i(t)C_i(t) + L_i'(t)\big)\Big)x(t)\\ &\quad \left. + (u(t) + K_i(t)x(t))'\big(R_i(t) + D_i'(t)P_i(t)D_i(t)\big)(u(t) + K_i(t)x(t))\big]dt|r_0 = i\right\}\\ &\quad + \mathbf{E}\{x'(T)(H_i - E'P_i(T)E)x(T)\}.\end{aligned} \tag{5.1.12}$$

Now, if $(P_1(\cdot),\ldots,P_l(\cdot)) \in C^1(0,T;\mathcal{S}_l^n)$ satisfy the CGDREs (5.1.9), then

$$\begin{aligned}J(x_0,i;u(\cdot)) &= x_0'E'P_i(0)Ex_0 + \mathbf{E}\left\{\int_0^T [u(t) + K_i(t)x(t)]'\big[R_i(t) + D_i'(t)P_i(t)D_i(t)\big]\right.\\ &\qquad \left. \times [u(t) + K_i(t)x(t)]dt|r_0 = i\right\} \geq x_0'E'P_i(0)Ex_0 > -\infty.\end{aligned} \tag{5.1.13}$$

Therefore, the LQ problem (5.1.6)–(5.1.8) is well-posed, and $J(x_0, i; u(\cdot))$ is minimized by the control given by (5.1.10) with the optimal value being $x_0' E' P_i(0) E x_0$. This completes the proof. □

5.1.2.2 Infinite-Time Horizon LQ Problem

Consider the following continuous-time stochastic Markov jump singular systems with state- and control-dependent noise

$$\begin{cases} Edx(t) = [A(r_t)x(t) + B(r_t)u(t)]dt + [C(r_t)x(t) + D(r_t)u(t)]dw(t), \\ x(0) = x_0 \in \mathbb{R}^n, \end{cases} \tag{5.1.14}$$

where $A(r_t) = A_i$, $B(r_t) = B_i$, $C(r_t) = C_i$ and $D(r_t) = D_i$ when $r_t = i$, while A_i etc., $i = 1, \ldots, l$, are given matrices of suitable sizes. $w(t)$ is a given one-dimensional standard Brownian motion on $[0, \infty)$, and a process $u(\cdot)$ is called the control input if $u(\cdot) \in L^2_{\mathcal{F}}(\mathbb{R}^m)$.

Due to the problem is considerated in infinite-time horizon, we introduce the concept of mean-square stable.

Definition 5.1.4 A control $u(\cdot)$ is called mean square stabilizing with respect to a given initial state (x_0, i) if the corresponding state $x(\cdot)$ of (5.1.14) with $x(0) = x_0$ and $r_0 = i$ satisfies $\lim_{t \to \infty} \mathbf{E}\|x(t)\|^2 = 0$.

Definition 5.1.5 The system (5.1.14) is called mean square stabilizable if there exists a feedback control $u^*(t) = \sum_{i=1}^{l} K_i \chi_{r_t = i}(t) x(t)$, where $K_1, \ldots, K_l$ are given matrices, which is stabilizing with respect to any initial state (x_0, i).

Next, for a given $(x_0, i) \in \mathbb{R}^n \times \Xi$, we define the corresponding set of admissible controls:

$$\mathcal{U}(x_0, i) = \left\{ u(\cdot) \in L^2_{\mathcal{F}}(\mathbb{R}^m) | u(\cdot) \, \textit{is mean square stabilizing with respect to} \, (x_0, i) \right\}.$$

For each $(x_0, i, u(\cdot)) \in \mathbb{R}^n \times \Xi \times \mathcal{U}(x_0, i)$, the LQ problem is to find a control $u(\cdot) \in \mathcal{U}(x_0, i)$ which minimizes the following quadratic cost associated with (5.1.14)

$$J(x_0, i; u(\cdot)) = \mathbf{E}\left\{ \int_0^\infty \begin{bmatrix} x(t) \\ u(t) \end{bmatrix}' \begin{bmatrix} Q(r_t) & L(r_t) \\ L'(r_t) & R(r_t) \end{bmatrix} \begin{bmatrix} x(t) \\ u(t) \end{bmatrix} dt | r_0 = i \right\}, \tag{5.1.15}$$

where $Q(r_t) = Q_i$, $R(r_t) = R_i$ and $L(r_t) = L_i$ when $r_t = i$, while Q_i, etc., $i = 1, \ldots, l$, are given matrices with suitable sizes. The value function V is defined as

$$V(x_0, i) = \inf_{u(\cdot)\in\mathcal{U}(x_0,i)} J(x_0, i; u(\cdot)). \tag{5.1.16}$$

Definition 5.1.6 The LQ problem (5.1.14)–(5.1.16) is called well-posed if

$$-\infty < V(x_0, i) < +\infty, \quad \forall (x_0, i) \in \mathbb{R}^n \times \Xi.$$

A well-posed problem is called attainable (with respect to (x_0, i)) if there is a control $u^*(\cdot) \in \mathcal{U}(x_0, i)$ that achieves $V(x_0, i)$. In this case the control $u^*(\cdot)$ is called optimal (with respect to (x_0, i)).

The following two basic assumptions are imposed in this section.

Assumption 5.2 The system (5.1.14) is mean square stabilizable.

Assumption 5.3 The data appearing in the LQ problem (5.1.14)–(5.1.16) satisfy, for every i,

$$A_i, C_i \in \mathbb{R}^{n\times n}, B_i, D_i \in \mathbb{R}^{n\times m}, Q_i \in \mathcal{S}^n, L_i \in \mathbb{R}^{n\times m}, R_i \in \mathcal{S}^m.$$

Before we give the main results of the infinite-time horizon LQ problem, we first present a technical lemma which is useful in our subsequent analysis.

Lemma 5.1.4 *Let matrices* $(P_1, \ldots, P_l) \in \mathcal{S}_l^n$ *be given, and* $P(r_t) = P_i$ *while* $r_t = i$. *Then for any admissible pair* $(x(\cdot), u(\cdot))$ *of the system* (5.1.14), *we have*

$$\begin{aligned}
&\mathbf{E}\left\{\int_0^T \begin{bmatrix} x(t) \\ u(t) \end{bmatrix}' M(P(r_t)) \begin{bmatrix} x(t) \\ u(t) \end{bmatrix} dt | r_0 = i\right\} \\
&= \mathbf{E}\left[x'(T)E'P(r_T)Ex(T) - x_0'E'P(r_0)Ex_0 | r_0 = i\right],
\end{aligned} \tag{5.1.17}$$

where

$$M(P_i) = \begin{bmatrix} \begin{array}{c} E'P_iA_i + A_i'P_iE + C_i'P_iC_i \\ + \sum_{j=1}^{l} \pi_{ij}E'P_jE \end{array} & E'P_iB_i + C_i'P_iD_i \\ B_i'P_iE + D_i'P_iC_i & D_i'P_iD_i \end{bmatrix}.$$

Proof Setting $\varphi(t, x, i) = x'E'P_iEx$ and applying the generalized Itô's formula to the linear system (5.1.14), we have

$$\begin{aligned}
&\mathbf{E}\left[x'(T)E'P(r_T)Ex(T) - x_0'E'(r_0)P(r_0)E(r_0)x_0 | r_0 = i\right] \\
&= \mathbf{E}[\varphi(T, x(T), r_T) - \varphi(0, x(0), r_0) | r_0 = i] \\
&= \mathbf{E}\left\{\int_0^T \Gamma\varphi(t, x(t), r_t)dt | r_0 = i\right\},
\end{aligned}$$

where

$$\begin{aligned}\Gamma\varphi(t,x,i) &= \varphi_t(t,x,i) + b'(t,x,u,i)\varphi_x(t,x,i) \\ &\quad + \frac{1}{2}tr[\sigma'(t,x,u,i)\varphi_{xx}(t,x,i)\sigma(t,x,u,i)] + \sum_{j=1}^{l}\pi_{ij}\varphi(t,x,j) \\ &= x'\left[E'P_iA_i + A_i'P_iE + C_i'P_iC_i + \sum_{j=1}^{l}\pi_{ij}E'P_jE\right]x + 2u'\left[B_i'P_iE_i + D_i'P_iC_i\right]x + u'D_i'P_iD_iu \\ &= \begin{bmatrix} x \\ u \end{bmatrix}' M(P(r_t))\begin{bmatrix} x \\ u \end{bmatrix}.\end{aligned}$$

This completes the proof. □

Theorem 5.1.2 *The infinite-time horizon LQ problem* (5.1.14)–(5.1.16) *is well-posed, if the following coupled generalized algebraic Riccati equations (CGAREs)* (5.1.18) *admit a solution* $(P_1, \ldots, P_l) \in \mathcal{S}_l^n$

$$\begin{cases} E'P_iA_i + A_i'P_iE + C_i'P_iC_i + Q_i + \sum_{j=1}^{l}\pi_{ij}E'P_jE \\ -\left(E'P_iB_i + C_i'P_iD_i + L_i\right)\left(R_i + D_i'P_iD_i\right)^{-1}\left(B_i'P_iE + D_i'P_iC_i + L_i'\right) = 0, \\ R_i + D_i'P_iD_i > 0, \quad i = 1, \ldots, l. \end{cases} \tag{5.1.18}$$

And the corresponding optimal feedback control law is

$$u^*(t) = \sum\nolimits_{i=1}^{l} K_i\chi_{r_t=i}(t)x(t), \tag{5.1.19}$$

where $K_i = -\left(R_i + D_i'P_iD_i\right)^{-1}\left(B_i'P_iE + D_i'P_iC_i + L_i'\right)$. *Furthermore, the cost corresponding to the control* $u^*(t) = \sum_{i=1}^{l} K_i\chi_{r_t=i}(t)x(t)$ *with the initial state* (x_0, i) *is*

$$J(x_0, i; u^*(\cdot)) = x_0'E'P_iEx_0, \quad i = 1, \ldots, l. \tag{5.1.20}$$

Proof By Lemma 5.1.4 and (5.1.18), using the square completion technique, we immediately have

$$\begin{aligned} J(x_0, i; u(\cdot)) &= x_0'E'P_iEx_0 + \mathbf{E}\left\{\int_0^{\infty}\begin{bmatrix} x(t) \\ u(t) \end{bmatrix}' M_1(P(r_t))\begin{bmatrix} x(t) \\ u(t) \end{bmatrix} dt | r_0 = i\right\} \\ &= x_0'E'P_iEx_0 + \mathbf{E}\left\{\int_0^{\infty} [u(t) + K_ix(t)]'H(P(r_t))[u(t) + K_ix(t)]dt | r_0 = i\right\} \\ &\geq J(x_0, i; u^*(\cdot)) = x_0'E'P_iEx_0, \quad i = 1, \ldots, l. \end{aligned} \tag{5.1.21}$$

where

$$M_1(P_i) = \begin{bmatrix} H_0(P_i) & H_1(P_i) \\ H_1'(P_i) & H(P_i) \end{bmatrix},\ H_0(P_i) = E'P_iA_i + A_i'P_iE + C_i'P_iC_i + Q_i + \sum_{j=1}^{l} \pi_{ij}E'P_jE,$$
$$H_1(P_i) = E'P_iB_i + C_i'P_iD_i + L_i,\ H(P_i) = R_i + D_i'P_iD_i.$$

From (5.1.21), we see that LQ problem (5.1.14)–(5.1.16) is well-posed, and $u^*(t) = \sum_{i=1}^{l} K_i\chi_{r_t=i}(t)x(t)$ is the optimal control. □

5.2 Two Person Zero-Sum Differential Games

5.2.1 Finite-Time Horizon Case

5.2.1.1 Problem Formulation

In this section, we will solve a two person zero-sum differential game on time interval $[0,T]$.

Consider the following controlled linear stochastic singular system

$$\begin{cases} Edx(t) = [A(t,r_t)x(t) + B_1(t,r_t)u_1(t) + B_2(t,r_t)u_2(t)]dt \\ \qquad\qquad + [C(t,r_t)x(t) + D_1(t,r_t)u_1(t) + D_2(t,r_t)u_2(t)]dw(t),\ t\in[0,T], \\ x(0) = x_0. \end{cases} \tag{5.2.1}$$

In the above $x(\cdot)\in\mathbb{R}^n$ is the state vector, $(x_0,r_0)\in\mathbb{R}^n\times\Xi$ is the initial value, $u_1(\cdot)$ and $u_2(\cdot)$ are two $\mathcal{F}_t$-adapted processes taking values in $\mathbb{R}^{m_1}$ and $\mathbb{R}^{m_2}$, which represent the controls of the two players, respectively. The admissible strategies of these two controls are denoted by $\mathcal{U}_1 \equiv L^2_{\mathcal{F}}(0,T;\mathbb{R}^{m_1})$ and $\mathcal{U}_2 \equiv L^2_{\mathcal{F}}(0,T;\mathbb{R}^{m_2})$. We assume that $A(t,r_t)$, $B_1(t,r_t)$, $B_2(t,r_t)$, $C(t,r_t)$, $D_1(t,r_t)$, and $D_2(t,r_t)$ are deterministic bounded matrix-valued functions of suitable sizes.

For every (x_0,i) and $(u_1(\cdot),u_2(\cdot))\in\mathcal{U}_1\times\mathcal{U}_2$, defining a quadratic cost function as following:

$$J(x_0,i;u_1(\cdot),u_2(\cdot)) = \mathbf{E}\left\{\int_0^T q(t,x(t),u_1(t),u_2(t),r_t)dt + x'(T)H(r_T)x(t)|r_0=i\right\}, \tag{5.2.2}$$

where

$$q(t,x,u_1,u_2,r_t) = \begin{bmatrix} x \\ u_1 \\ u_2 \end{bmatrix}' \begin{bmatrix} Q(t,r_t) & L_1'(t,r_t) & L_1'(t,r_t) \\ L_1(t,r_t) & R_1(t,r_t) & 0 \\ L_2(t,r_t) & 0 & R_2(t,r_t) \end{bmatrix} \begin{bmatrix} x \\ u_1 \\ u_2 \end{bmatrix},$$

with $Q(t,r_t)$, $L_1(t,r_t)$, $L_2(t,r_t)$, $R_1(t,r_t)$, $R_2(t,r_t)$ being deterministic bounded matrix-valued functions of suitable sizes and $H(r_T)$ being a given matrix.

In (5.2.1) and (5.2.2), $A(t,r_t) = A_i(t)$, etc. whenever $r_t = i$, and $H(r_T) = H_i$ whenever $r_T = i$.

Assumption 5.4 The data appearing in (5.2.1) and (5.2.2) satisfy, for every i

$$\begin{cases} A_i(\cdot),\ C_i(\cdot) \in L^\infty(0,\ T;\ \mathbb{R}^{n\times n}), \\ B_{1i}(\cdot),\ D_{1i}(\cdot) \in L^\infty(0,\ T;\ \mathbb{R}^{n\times m_1}), \\ B_{2i}(\cdot),\ D_{2i}(\cdot) \in L^\infty(0,\ T;\ \mathbb{R}^{n\times m_2}), \\ Q_i(\cdot) \in L^\infty(0,\ T;\ \mathcal{S}^n), \\ L_{1i}(\cdot) \in L^\infty(0,\ T;\ \mathbb{R}^{n\times m_1}), \\ L_{1i}(\cdot) \in L^\infty(0,\ T;\ \mathbb{R}^{n\times m_1}), \\ R_{1i}(\cdot) \in L^\infty(0,\ T;\ \mathcal{S}^{m_1}), \\ R_{2i}(\cdot) \in L^\infty(0,\ T;\ \mathcal{S}^{m_2}), \\ H_i \in \mathcal{S}^n. \end{cases}$$

In what follows, we denote

$$B(\cdot) = (B_1(\cdot),\ B_2(\cdot)),\ D(\cdot) = (D_1(\cdot),\ D_2(\cdot)),\ L(\cdot) = \begin{pmatrix} L_1(\cdot) \\ L_2(\cdot) \end{pmatrix},\ R(\cdot) = \begin{pmatrix} R_1(\cdot) & 0 \\ 0 & R_2(\cdot) \end{pmatrix}.$$

Our aim is to find a pair $(u_1^*(\cdot), u_2^*(\cdot)) \in \mathcal{U}_1 \times \mathcal{U}_2$ which is called an open-loop saddle point of the game over $[0,T]$ if the following inequality holds:

$$J(x_0,i;u_1^*(\cdot),u_2(\cdot)) \le J(x_0,i;u_1^*(\cdot),u_2^*(\cdot)) \le J(x_0,i;u_1(\cdot),u_2^*(\cdot)),\ \forall(u_1(\cdot),u_2(\cdot)) \in \mathcal{U}_1 \times \mathcal{U}_2. \tag{5.2.3}$$

5.2.1.2 Main Result

For the above game problem, if we take $V(t,x,i) = x'E'P_i(t)Ex$ with $P_i(\cdot)$ a symmetric matrix as the Lyapunov function candidate, and adopting the same procedure done in Sect. 5.1, we will have the following result.

Theorem 5.2.1 *For the game problem* (5.2.1)–(5.2.3), *if the following coupled generalized differential Riccati equations admit a solution* $(P_1(\cdot),\ldots,P_l(\cdot)) \in \mathcal{C}^1(0,T;\mathcal{S}_l^n)$ *(with the time argument t suppressed),*

$$\begin{cases} E'\dot{P}_iE + E'P_iA_i + A_i'P_iE + C_i'P_iC_i + Q_i + \sum_{j=1}^{l} \pi_{ij}E'P_jE \\ -(EPB_i + C_i'P_iD_i + L_i)\left(R_i + D_i'P_iD_i\right)^{-1}\left(B_i'P_iE + D_i'P_iC_i + L_i'\right) = 0, \\ E'P_i(T)E = H_i, \\ R_i + D_i'P_iD_i > 0, \end{cases} \tag{5.2.4}$$

where

$$B_i = (B_{1i},\ B_{2i}),\ D_i = (D_{1i},\ D_{2i}),\ L_i = \begin{pmatrix} L_{1i} \\ L_{2i} \end{pmatrix},\ R_i = \begin{pmatrix} R_{1i} & 0 \\ 0 & R_{2i} \end{pmatrix}.$$

Then an open-loop saddle point $u^*(\cdot) = \begin{pmatrix} u_1^*(\cdot) \\ u_2^*(\cdot) \end{pmatrix}$ *is*

$$u^*(t) = \sum_{i=1}^{l} K_i(t)\chi_{r_t=i}(t)x(t), \tag{5.2.5}$$

where $K_i(t) = -\left[R_i(t) + D_i'(t)P_i(t)D_i(t)\right]^{-1}\left[B_i'(t)P_i(t)E + D'(t)_iP_i(t)C_i(t) + L_i'(t)\right]$. *Moreover, the optimal value being* $x_0'E'P_i(0)Ex_0,\ i = 1,\ldots,l$.

5.2.2 *Infinite-Time Horizon Case*

5.2.2.1 Problem Formulation

In this subsection, we consider the two person zero-sum stochastic differential games on time interval $[0,\infty)$. Firstly, we define the following space

$L_2^{loc}(\mathbb{R}^m) := \{\phi(\cdot,\cdot) : [0,\infty) \times \Omega \to \mathbb{R}^m | \phi(\cdot,\cdot)$ is $\mathcal{F}_t$-adapted, Lebesgue measurable, and $\mathbf{E}\int_0^T \|\phi(t,\omega)\|^2 dt < \infty,\ \forall T > 0\}$.

Consider the following controlled Markov jump linear systems

$$\begin{cases} Edx(t) = [A(r_t)x(t) + B_1(r_t)u_1(t) + B_2(r_t)u_2(t)]dt + \\ \qquad [C(r_t)x(t) + D_1(r_t)u_1(t) + D_2(r_t)u_2(t)]dw(t), \\ x(0) = x_0, \end{cases} \tag{5.2.6}$$

where $x(\cdot) \in \mathbb{R}^n$ is the system state, $(x_0, r_0) \in \mathbb{R}^n \times \Xi$ is the initial state, $E \in \mathbb{R}^{n\times n}$ is a known singular matrix with $0 < rank(E) = k \leq n$, $A(r_t) = A_i$, $B_1(r_t) = B_{1i}$, $B_2(r_t) = B_{2i}$, $C(r_t) = C_i$, $D_1(r_t) = D_{1i}$ and $D_2(r_t) = D_{2i}$, when $r_t = i$,$i = 1, \ldots, l$, while $A(i)$, etc., are given matrices of suitable sizes. $u_1(\cdot) \in \mathcal{U}_1 \equiv L_2^{loc}(\mathbb{R}^{m_1})$ and $u_2(\cdot) \in \mathcal{U}_2 \equiv L_2^{loc}(\mathbb{R}^{m_2})$ are two admissible control processes, which represents the control strategies of these two players.

For system (5.2.6) and $(x_0, i) \in \mathbb{R}^n \times \Xi$, the corresponding sets of admissible controls are denoted by:

$\bar{\mathcal{U}}(x_0, i) = \{(u_1(\cdot), u_2(\cdot)) \in \mathcal{U}_1 \times \mathcal{U}_2 | (u_1(\cdot), u_2(\cdot))$ is mean-square stabilizing w.r.t. $(x_0, i)\}$.

For each (x_0, i) and $(u_1(\cdot), u_2(\cdot)) \in \bar{\mathcal{U}}(x_0, i)$, the cost function is

$$J(x_0, i; u_1(\cdot), u_2(\cdot)) = \mathbf{E}\left\{\int_0^\infty q(t, x(t), u_1(t), u_2(t), r_t)dt | r_0 = i\right\}, \tag{5.2.7}$$

where

$$q(t, x, u_1, u_2, r_t) = \begin{bmatrix} x \\ u_1 \\ u_2 \end{bmatrix}' \begin{bmatrix} Q(r_t) & L_1'(r_t) & L_2'(r_t) \\ L_1(r_t) & R_1(r_t) & 0 \\ L_2(r_t) & 0 & R_2(r_t) \end{bmatrix} \begin{bmatrix} x \\ u_1 \\ u_2 \end{bmatrix},$$

with $Q(r_t)$, $L_1(r_t)$, $L_2(r_t)$, $R_1(r_t)$ and $R_2(r_t)$ been given matrices with suitable sizes.

The problem is to look for $(u_1^*(\cdot), u_2^*(\cdot)) \in \bar{\mathcal{U}}(x_0, i)$ which is called the saddle point equilibrium for the game, such that

$$J(x_0, i; u_1^*(\cdot), u_2(\cdot)) \leq J(x_0, i; u_1^*(\cdot), u_2^*(\cdot)) \leq J(x_0, i; u_1(\cdot), u_2^*(\cdot)), \quad i = 1, \ldots, l. \tag{5.2.8}$$

5.2.2.2 Main Result

Mean-square stabilizability is a standard assumption in an infinite-horizon LQ control problem. So we use this assumption here.

Assumption 5.2.1 The system (5.2.6) is mean-square stabilizable.

Similar to the finite-time horizon two person zero-sum stochastic games discussed in last subsection, we can get the corresponding results of the infinite-time horizon two person zero-sum stochastic games stated as Theorem 5.2.2.

Theorem 5.2.2 *Suppose Assumption* 5.2.1 *holds, for the two person zero-sum stochastic games* (5.2.6)–(5.2.8) *and* $(x_0, i) \in \mathbb{R}^n \times \Xi$, *the strategy set* $(u_1^*(\cdot), u_2^*(\cdot))$ *is the equilibrium strategy if and only if the following algebraic Riccati equations*

$$\begin{cases} E'P_iA_i + A_i'P_iE + C_i'P_iC_i + Q_i + \sum_{j=1}^{l} \pi_{ij}E'P_jE - \left(EPB_i + C_i'P_iD_i + L_i\right) \\ \times \left(R_i + D_i'P_iD_i\right)^{-1}\left(B_i'P_iE + D_i'P_iC_i + L_i'\right) = 0, \\ R_i + D_i'P_iD_i > 0, \end{cases} \tag{5.2.9}$$

where

$$B_i = (B_{1i},\ B_{2i}),\ D_i = (D_{1i},\ D_{2i}),\ L_i = \begin{pmatrix} L_{1i} \\ L_{2i} \end{pmatrix},\ R_i = \begin{pmatrix} R_{1i} & 0 \\ 0 & R_{2i} \end{pmatrix}.$$

admit a solution $(P_1, \ldots, P_l) \in \mathcal{S}_l^n$. *In this case,* $u^*(\cdot) = \begin{pmatrix} u_1^*(\cdot) \\ u_2^*(\cdot) \end{pmatrix}$ *can be represented as*

$$u^*(t) = \sum_{i=1}^{l} K_i \chi_{r_t=i}(t)x(t), \tag{5.2.10}$$

where $K_i = -\left[R_i + D_i'P_iD_i\right]^{-1}\left[B_i'P_iE + D_i'P_iC_i + L_i'\right]$. *Moreover, the optimal value is* $x_0'E'P_iEx_0$, $i = 1, \ldots, l$.

5.3 Stochastic Nash Differential Games with Two Player

5.3.1 Finite-Time Horizon Case

5.3.1.1 Problem Formulation

First, we consider a stochastic Nash differential game with two player on a finite horizon $[0, T]$, N-player case is similar.

Consider the following Markov jump singular systems described by stochastic differential equation

$$\begin{cases} Edx(t) = [A(t, r_t)x(t) + B_1(t, r_t)u(t) + B_2(t, r_t)v(t)]dt \\ \qquad\qquad + [C(t, r_t)x(t) + D_1(t, r_t)u(t) + D_2(t, r_t)v(t)]dw(t), \\ x(0) = x_0 \in \mathbb{R}^n. \end{cases} \tag{5.3.1}$$

where $E \in \mathbb{R}^{n\times n}$ is a known singular matrix with $0<rank(E)=k\leq n$, $x(\cdot)\in\mathbb{R}^n$ is the system state, $(x_0,r_0)\in\mathbb{R}^n\times\Xi$ is the initial state, two admissible controls $u(\cdot)$ and $v(\cdot)$ are $\mathcal{F}_t$-adapted, $\mathbb{R}^{n_u}$- and $\mathbb{R}^{n_v}$-valued measurable process on $[0,T]$. The sets of all admissible controls are denoted by $\mathcal{U}\equiv L^2_{\mathcal{F}}(0,T;\mathbb{R}^{n_u})$ and $\mathcal{V}\equiv L^2_{\mathcal{F}}(0,T;\mathbb{R}^{n_v})$.

For each $(x_0,r_0)\in\mathbb{R}^n\times\Xi$ and $(u(\cdot),v(\cdot))\in\mathcal{U}\times\mathcal{V}$, the cost function $J_k(x_0,i;u(\cdot),v(\cdot))$ is

$$J_k(x_0,i;u(\cdot),v(\cdot))=\mathbf{E}\left\{\int_0^T z'(t)M_k(t,r_t)z(t)dt+x'(T)H_k(r_T)x(T)|r_0=i\right\},$$
$$z(t)=\begin{bmatrix}x(t)\\u(t)\\v(t)\end{bmatrix},M_k(t,r_t)=\begin{bmatrix}Q_k(t,r_t)&L_{k1}(t,r_t)&L_{k2}(t,r_t)\\L'_{k1}(t,r_t)&R_{k1}(t,r_t)&0\\L'_{k1}(t,r_t)&0&R_{k2}(t,r_t)\end{bmatrix},k=1,\ 2. \tag{5.3.2}$$

In (5.3.1) and (5.3.2), $A(t,r_t)=A_i(t)$, etc. whenever $r_t=i$, and $H_k(r_T)=H_{ki}$, $k=1,\ 2$, whenever $r_T=i$.

Assumption 5.3.1 The data appearing in the finite horizon stochastic Nash differential game problem (5.3.1)–(5.3.2) satisfy, for every i,

$$\begin{cases}A_i(\cdot),C_i(\cdot)\in L^\infty(0,T;\mathbb{R}^n), & B_{1i}(\cdot),D_{1i}(\cdot)\in L^\infty(0,T;\mathbb{R}^{n\times n_u}),\\ B_{2i}(\cdot),D_{2i}(\cdot)\in L^\infty(0,T;\mathbb{R}^{n\times n_v}), & Q_{1i}(\cdot)\in L^\infty(0,T;\mathcal{S}^n),\\ Q_{2i}(\cdot)\in L^\infty(0,T;\mathcal{S}^n), & R_{11i}(\cdot)\in L^\infty(0,T;\mathcal{S}^{n_u}),\\ L_{11i}(\cdot)\in L^\infty(0,T;\mathbb{R}^{n\times n_u}), & L_{12i}(\cdot)\in L^\infty(0,T;\mathbb{R}^{n\times n_v}),\\ L_{21i}(\cdot)\in L^\infty(0,T;\mathbb{R}^{n\times n_u}), & L_{22i}(\cdot)\in L^\infty(0,T;\mathbb{R}^{n\times n_v}),\\ R_{22i}(\cdot)\in L^\infty(0,T;\mathcal{S}^{n_v}), & H_{1i}\in\mathcal{S}^n,H_{2i}\in\mathcal{S}^n.\end{cases}$$

Now, let's give the form definition of finite time stochastic Nash differential games:

Definition 5.3.1 For each $(x_0,r_0)\in\mathbb{R}^n\times\Xi$, finding an admissible control pair $(u^*(\cdot),v^*(\cdot))\in\mathcal{U}\times\mathcal{V}$ which is called the Nash equilibrium for the game, such that

$$\begin{cases}J_1(x_0,i;u^*(\cdot),v^*(\cdot))\leq J_1(x_0,i;u(\cdot),v^*(\cdot)), & \forall u(\cdot)\in\mathcal{U},\\ J_2(x_0,i;u^*(\cdot),v^*(\cdot))\leq J_2(x_0,i;u^*(\cdot),v(\cdot)), & \forall v(\cdot)\in\mathcal{V}.\end{cases} \tag{5.3.3}$$

5.3.1.2 Main Result

With the help of the relevant conclusions of differential game with one person, it is easy to obtain the following conclusions:

Theorem 5.3.1 *For the finite time stochastic Nash differential game* (5.3.1)–(5.3.2), *there exists the Nash equilibrium* $(u^*(\cdot),v^*(\cdot))$, *if and only if the following coupled generalized differential Riccati equations (with time t supressed)*

$$\begin{cases} E'\dot{P}_{1i}E + E'P_{1i}\bar{A}_i + \bar{A}_i'P_{1i}E' + \bar{C}_i'P_{1i}\bar{C}_i + \bar{Q}_{1i} + \sum_{j=1}^{l}\pi_{ij}E'P_{1j}E \\ -\left(E'P_{1i}B_{1i} + \bar{C}_i'P_{1i}D_{1i} + L_{11i}\right)\left(R_{11i} + D_{1i}'P_{1i}D_{1i}\right)^{-1}\left(B_{1i}'P_{1i}E + D_{1i}'P_{1i}\bar{C}_i + L_{11i}'\right) = 0, \\ E'P_{1i}(T)E = H_{1i}, \\ R_{11i} + D_{1i}'P_{1i}D_{1i} > 0, i = 1, \ldots, l. \end{cases} \tag{5.3.4}$$

$$K_{1i} = -\left(R_{11i} + D_{1i}'P_{1i}D_{1i}\right)^{-1}\left(B_{1i}'P_{1i}E + D_{1i}'P_{1i}\bar{C}_i + L_{11i}'\right), \tag{5.3.5}$$

$$\begin{cases} E'\dot{P}_{2j}E + E'P_{2j}\tilde{A}_j + \tilde{A}_j'P_{2j}E + \tilde{C}_j'P_{2j}\tilde{C}_j + \tilde{Q}_{2j} + \sum_{k=1}^{l}\pi_{jk}E'P_{2k}E \\ -\left(E'P_{2j}B_{2j} + \tilde{C}_j'P_{2j}D_{2j} + L_{22j}\right)\left(R_{22j} + D_{2j}'P_{2j}D_{2j}\right)^{-1}\left(B_{2j}'P_{2j}E + D_{2j}'P_{2j}\tilde{C}_j + L_{2j}'\right) = 0, \\ E'P_{2j}(T)E = H_{2j}, \\ R_{22j}(j) + D_{2j}'(j)P_{2j}(j)D_{2j}(j) > 0, j = 1, \ldots, l. \end{cases} \tag{5.3.6}$$

$$K_{2j} = -\left(R_{22j} + D_{2j}'P_{2j}D_{2j}\right)^{-1}\left(B_{2j}'P_{2j}E + D_{2j}'P_{2j}\tilde{C}_j + L_{22j}'\right). \tag{5.3.7}$$

where

$$\bar{A}_i = A_i + B_{2i}K_{2i},\ \bar{C}_i = C_i + D_{2i}K_{2i},\ \bar{Q}_{1i} = Q_{1i} + L_{12i}K_{2i} + K_{2i}'L_{12i}' + K_{2i}'R_{12i}K_{2i},$$
$$\tilde{A}_j = A_j + B_{1j}K_{1j},\ \tilde{C}_j = C_j + D_{1j}K_{1j},\ \tilde{Q}_{2j} = Q_{2j} + L_{21j}K_{1j} + K_{1j}'L_{21j}' + K_{1j}'R_{21j}K_{1j}.$$

admit a solution $P(\cdot) = (P_1(\cdot), P_2(\cdot))$ *with* $P_1(\cdot) = (P_{11}(\cdot), \ldots, P_{1l}(\cdot)) \in \mathcal{C}^1(0, T; \mathcal{S}_l^n) \geq 0$, $P_2(\cdot) = (P_{21}(\cdot), \ldots, P_{2l}(\cdot)) \in \mathcal{C}^1(0, T; \mathcal{S}_l^n) \geq 0$.

Denote $F_{1i}^*(t) = K_{1i}(t)$, $F_{2i}^*(t) = K_{2i}(t)$, *then the Nash equilibrium strategy* $(u^*(\cdot), v^*(\cdot))$ *can be represented by*

$$\begin{cases} u^*(t) = \sum_{i=1}^{l} F_{1i}^*(t)\chi_{r_t=i}(t)x(t), \\ v^*(t) = \sum_{i=1}^{l} F_{2i}^*(t)\chi_{r_t=i}(t)x(t). \end{cases}$$

Moreover, the optimal value is

$$J_k(x_0, i; u^*(\cdot), v^*(\cdot)) = x_0'E'P_{ki}(0)Ex_0, \quad k = 1, 2, \quad i = 1, \ldots, l.$$

Proof These results can be proved by using the concept of Nash equilibrium described in Definition 5.3.1 as follows. Given $v^*(t) = \sum_{i=1}^{l} F_{2i}^*(t)\chi_{r_t=i}(t)x(t)$ is the optimal control strategy implemented by player P_2, player P_1 facing the following optimization problem:

$$\min_{u(\cdot)\in\mathcal{U}} \mathbf{E}\left\{\int_0^T \begin{bmatrix} x(t) \\ u(t) \end{bmatrix}' \begin{bmatrix} \bar{Q}_1(t,r_t) & L_{11}(t,r_t) \\ L_{11}'(t,r_t) & R_{11}(t,r_t) \end{bmatrix} \begin{bmatrix} x(t) \\ u(t) \end{bmatrix} dt + x'(T)H_1(r_T)x(T)|r_0 = i\right\},$$

$s.t.$

$$\begin{cases} Edx(t) = [\bar{A}(t,r_t)x(t) + B_1(t,r_t)u(t)]dt + [\bar{C}(t,r_t)x(t) + D_1(t,r_t)u(t)]dw(t), \\ x(0) = x_0. \end{cases} \tag{5.3.8}$$

where $\bar{Q}_1 = Q_1 + (F_2^*)'L_{12}' + L_{12}F_2^* + (F_2^*)'R_{12}F_2^*$.

Note that the above optimization problem defined in (5.3.8) is a standard stochastic LQ problem. Applying Theorem 5.1.1 to this optimization problem as

$$\begin{bmatrix} \bar{Q}_1(r_t) & L_{11}(r_t) \\ L_{11}'(r_t) & R_{11}(r_t) \end{bmatrix} \Rightarrow \begin{bmatrix} Q_1 & L_1 \\ L_1' & R_{11} \end{bmatrix}, \bar{A} \Rightarrow A,\ \bar{C} \Rightarrow C.$$

We can easily get the optimal control and the optimal value function

$$u^*(t) = \sum_{i=1}^{l} F_{1i}^*(t)x(t)\chi_{\{r_t=i\}}(t), J_1(x_0, i; u^*(\cdot), v^*(\cdot)) = x_0'E'P_{1i}(0)x_0,\ i = 1, \ldots, l. \tag{5.3.9}$$

Similarly, we can prove that $v^*(t) = \sum_{i=1}^{l} F_{2i}^*(t)\chi_{r_t=i}(t)x(t)$ is the optimal control strategy of player P_2.

This completes the proof of Theorem 5.3.1. □

5.3.2 *Infinite-Time Horizon Case*

5.3.2.1 Problem Formulation

In this subsection, we discuss the stochastic Nash differential games on time interval $[0, \infty)$. Before giving the problem to be discussed, first define the following space

$L_2^{loc}(\mathbb{R}^m) := \{\phi(\cdot,\cdot) : [0,\infty) \times \Omega \to \mathbb{R}^m | \phi(\cdot,\cdot)$ is $\mathcal{F}_t$-adapted, Lebesgue measurable, and $\mathbf{E}\int_0^T \|\phi(t,\omega)\|^2 dt < \infty,\ \forall T > 0\}$.

Consider the following Markov jump singular systems defined by

$$\begin{cases} Edx(t) = [A(r_t)x(t) + B_1(r_t)u(t) + B_2(r_t)v(t)]dt \\ \qquad\qquad + [C(r_t)x(t) + D_1(r_t)u(t) + D_2(r_t)v(t)]dw(t), \\ x(0) = x_0. \end{cases} \tag{5.3.10}$$

where $A(r_t) = A(i)$, $B_1(r_t) = B_1(i)$, $B_2(r_t) = B_2(i)$, $C(r_t) = C(i)$, $D_1(r_t) = D_1(i)$ and $D_2(r_t) = D_2(i)$, when $r_t = i$, $i = 1,\ldots,l$, while $A(i)$, etc., are given matrices of suitable sizes. $u(\cdot) \in \mathcal{U} \equiv L_2^{loc}(\mathbb{R}^{n_u})$ and $v(\cdot) \in \mathcal{V} \equiv L_2^{loc}(\mathbb{R}^{n_v})$ are two admissible control processes, which represents the control strategies of these two players.

Next, for a given $(x_0, i) \in \mathbb{R}^n \times \Xi$, we define the corresponding sets of admissible controls:

$\bar{\mathcal{U}}(x_0, i) = \{(u(\cdot), v(\cdot)) \in \mathcal{U} \times \mathcal{V} | (u(\cdot), v(\cdot))$ is mean-square stabilizing w.r.t. $(x_0, i)\}$.

For each (x_0, i) and $(u(\cdot), v(\cdot)) \in \bar{\mathcal{U}}(x_0, i)$, the cost function $J_k(x_0, i; u(\cdot), v(\cdot))$ is

$$\begin{aligned} J_k(x_0, i; u(\cdot), v(\cdot)) &= \mathbf{E}\left\{\int_0^\infty z'(t)M_k(r_t)z(t)dt | r_0 = i\right\}, \\ z(t) = \begin{bmatrix} x(t) \\ u(t) \\ v(t) \end{bmatrix}, M_k(r_t) &= \begin{bmatrix} Q_k(r_t) & L_{k1}(r_t) & L_{k2}(r_t) \\ L'_{k1}(r_t) & R_{k1}(r_t) & 0 \\ L'_{k2}(r_t) & 0 & R_{k2}(r_t) \end{bmatrix}, k = 1, 2. \end{aligned} \tag{5.3.11}$$

In (5.3.10) and (5.3.11), $A(r_t) = A(i)$, ..., when $r_t = i$, while $A(i)$, etc., are given matrices with suitable sizes.

The form definition of infinite-time horizon stochastic Nash differential game is given below:

Definition 5.3.2 For each $(x_0, i) \in \mathbb{R}^n \times \Xi$, finding an admissible control pair $(u^*(\cdot), v^*(\cdot)) \in \bar{\mathcal{U}}(x_0, i)$ which is called the Nash equilibrium for the game, such that,

$$\begin{cases} J_1(x_0, i; u^*(\cdot), v^*(\cdot)) \le J_1(x_0, i; u(\cdot), v^*(\cdot)), & \forall u(\cdot) \in \mathcal{U}, \\ J_2(x_0, i; u^*(\cdot), v^*(\cdot)) \le J_2(x_0, i; u^*(\cdot), v(\cdot)), & \forall v(\cdot) \in \mathcal{V}. \end{cases} \tag{5.3.12}$$

5.3.2.2 Main Result

Firstly, we give a standard assumption used in stochastic LQ problems.

Assumption 5.3.1 The system (5.3.10) is mean-square stabilizable.

Similar to the finite-time horizon stochastic Nash games discussed in last subsection, we can get the corresponding results of the infinite-time horizon stochastic Nash games stated as Theorem 5.3.2.

Theorem 5.3.2 *Suppose Assumption* 5.3.1 *holds, the infinite-time horizon stochastic Nash differential game* (5.3.10)–(5.3.11) *has a Nash equilibrium* $(u^*(\cdot), v^*(\cdot))$, *if and only if the following algebraic Riccati equations admit a solution* $P = (P_1, P_2) \in \mathcal{S}_n^l \times \mathcal{S}_n^l \geq 0$ *with* $P_1 = (P_1(1), \ldots, P_1(l))$, $P_2 = (P_2(1), \ldots, P_2(l))$:

$$\begin{cases} E'P_1(i)\bar{A}(i) + \bar{A}'(i)P_1(i)E + \bar{C}_1^i(i)P_1(i)\bar{C}_1(i) + \bar{Q}_1(i) + \sum_{j=1}^{l} \pi_{ij}E'P_1(j)E \\ -\left(E'P_1(i)B_1(i) + \bar{C}_1'(i)P_1(i)D_1(i) + L_{11}(i)\right)\left(R_{11}(i) + D_1'(i)P_1(i)D_1(i)\right)^{-1} \\ \times\left(B_1'(i)P_1(i)E + D_1'(i)P_1(i)\bar{C}_1(i) + L_{11}'(i)\right) = 0, \\ R_{11}(i) + D_1'(i)P_1(i)D_1(i) > 0, \ i \in \Xi, \end{cases} \tag{5.3.13}$$

$$K_1 = -\left(R_{11}(i) + D_1'(i)P_1(i)D_1(i)\right)^{-1}\left(B_1'(i)P_1(i)E + D_1'(i)P_1(i)\bar{C}_1(i) + L_{11}'(i)\right), \tag{5.3.14}$$

$$\begin{cases} E'P_2(j)\tilde{A}(j) + \tilde{A}'(j)P_2(j)E + \tilde{C}_2'(j)P_2(j)\tilde{C}_2(j) + \tilde{Q}_2(j) + \sum_{k=1}^{l} \pi_{jk}E'P_2(k)E \\ -\left(E'P_2(j)B_2(j) + \tilde{C}_2'(j)P_2(j)D_2(j) + L_{22}(j)\right)\left(R_{22}(j) + D_2'(j)P_2(j)D_2(j)\right)^{-1} \\ \times\left(B_2'(j)P_2(j)E + D_2'(j)P_2(j)\tilde{C}_2(j) + L_{22}'(j)\right) = 0, \\ R_{22}(j) + D_2'(j)P_2(j)D_2(j) > 0, \ j \in \Xi, \end{cases} \tag{5.3.15}$$

$$K_2 = -\left(R_{22}(j) + D_2'(j)P_2(j)D_2(j)\right)^{-1}\left(B_2'(j)P_2(j)E + D_2'(j)P_2(j)\tilde{C}_2(j) + L_{22}'(j)\right), \tag{5.3.16}$$

where

$$\begin{aligned} \bar{A} &= A + B_2K_2, \quad \bar{C}_1 = C + D_2K_2, \quad \bar{Q}_1 = Q_1 + L_{12}K_2 + K_2'L_{12}' + K_2'R_{12}K_2, \\ \tilde{A} &= A + B_1K_1, \quad \tilde{C}_2 = C + D_1K_1, \quad \tilde{Q}_2 = Q_2 + L_{21}K_1 + K_1'L_{21}' + K_1'R_{21}K_1. \end{aligned}$$

In this case, the equilibrium strategies and optimal cost function are

$$\begin{aligned} &u^*(t) = \sum_{i=1}^{l} K_1(i)\chi_{r_t=i}(t)x(t), \ v^*(t) = \sum_{i=1}^{l} K_2(i)\chi_{r_t=i}(t)x(t), \\ &J_k(x_0, i; u^*(\cdot), v^*(\cdot)) = x_0'E'P_k(i)Ex_0, \quad k = 1, 2, \quad i = 1, \ldots, l. \end{aligned}$$

5.4 Stochastic Stackelberg Differential Game with Two Person

5.4.1 Problem Formulation

Consider the following Markov jump singular systems described by stochastic differential equation

$$\begin{cases} Edx(t) = [A(r_t)x(t) + B_1(r_t)u_1(t) + B_2(r_t)u_2(t)]dt + C(r_t)x(t)dw(t), \\ x(0) = x_0. \end{cases} \tag{5.4.1}$$

where $x(t) \in \mathbb{R}^n$ represents the system state, $u_k(t) \in \mathbb{R}^{m_k}$, $k = 1, 2$ represent the k-th control inputs, $E \in \mathbb{R}^{n \times n}$ is a known singular matrix with $0 < rank(E) = k \leq n$, $(x_0, r_0) \in \mathbb{R}^n \times \Xi$ is the initial state. It is assumed that the player denoted by u_2 is the leader and the player denoted by u_1 is the follower. In (5.4.1), $A(r_t) = A(i)$, $B_k(r_t) = B_k(i)$, $k = 1, 2$, $C(r_t) = C(i)$, when $r_t = i$, $i = 1, \ldots, l$, while $A(i)$, etc., are given matrices of suitable sizes.

Without loss of generality, the stochastic dynamic games are investigated under the following basic assumption:

Assumption 5.4.1 (A, B_k, C), $k = 1, 2$ is stabilizable.

For each initial value (x_0, i), the cost function for each strategy subset is defined by

$$J_k(x_0, i; u_1, u_2) = \mathbf{E}\left\{ \int_0^\infty \begin{bmatrix} x'(t)Q_k(r_t)x(t) + u_k'(t)R_{kk}(r_t)u_k(t) \\ + u_j'(t)R_{kj}(r_t)u_j(t) \end{bmatrix} dt | r_0 = i \right\}, \tag{5.4.2}$$

where $k = 1, 2$, $Q_k(r_t) = Q_k'(r_t) \geq 0$, $R_{kk}(r_t) = R_{kk}'(r_t) > 0$, $R_{kj}(r_t) = R_{kj}'(r_t) \geq 0$, $k \neq j$.

5.4.2 Main Result

Without loss of generality, we restrict the control strategy of each player as linear state feedback case, i.e., the closed-loop Stackelberg strategies $u_k(t) = u_k(x, t)$ have the following form

$$u_k(t) = \sum_{i=1}^{l} F_k(i)\chi_{r_t = i}(t)x(t).$$

The Stackelberg strategy of the game system (5.4.1)–(5.4.2) is defined as:

Definition 5.4.1 [9] a strategy set (u_1^*, u_2^*) is called a Stackelberg strategy if the following conditions hold

$$J_2(x_0, i; u_1^*, u_2^*) \le J_2(x_0, i; u_1^o(u_2), u_2), \quad \forall u_2 \in \mathbb{R}^{m_2}, \tag{5.4.3}$$

where

$$J_1(x_0, i; u_1^o(u_2), u_2) = \min_v J_1(x_0, i; u_1, u_2), \tag{5.4.4}$$

and

$$u_1^* = u_1^0(u_2^*). \tag{5.4.5}$$

Theorem 5.4.1 *Suppose that the following cross-coupled algebraic matrix Eqs.* (5.4.6a–5.4.6e) *has solutions* $\bar{M}_k(i) \ge 0$, $\bar{N}_k(i)$, $k = 1, 2$ *and* $F_2(i)$

$$\begin{aligned} &A'_{F_1}(i)\bar{M}_1(i)E + E'\bar{M}_1(i)A_{F_1}(i) + C'(i)\bar{M}_1(i)C(i) + Q_{F_1}(i) \\ &- F'_1(i)R_{11}(i)F_1(i) + \sum_{j=1}^{l} \pi_{ij}E'\bar{M}_1(j)E = 0, \end{aligned} \tag{5.4.6a}$$

$$\begin{aligned} &A'_F(i)\bar{M}_2(i)E + E'\bar{M}_2(i)A_F(i) + C'(i)\bar{M}_2(i)C(i) + Q_{F_2}(i) \\ &+ F'_1(i)R_{21}(i)F_1(i) + \sum_{j=1}^{l} \pi_{ij}E'\bar{M}_2(j)E = 0, \end{aligned} \tag{5.4.6b}$$

$$\begin{aligned} &A_{F_1}(i)\bar{N}_1(i)E + E'\bar{N}_1(i)A'_{F_1}(i) + C(i)\bar{N}_1(i)C'(i) - E'B_1(i)R_{11}^{-1}(i)B'_1(i)\bar{M}_1(i)\bar{N}_1(i)E \\ &- E'\bar{N}_1(i)\bar{M}_1(i)B_1(i)R_{11}^{-1}(i)B'_1(i)E + \pi_{ii}E'\bar{N}_1(i)E - B_1(i)R_{11}^{-1}(i)B'_1(i)\bar{M}_2(i)E\bar{N}_2(i) \\ &- \bar{N}_2(i)E'\bar{M}_2(i)B_1(i)R_{11}^{-1}(i)B'_1(i) + E'B_1(i)R_{11}^{-1}(i)R_{21}(i)R_{11}^{-1}(i)B'_1(i)\bar{M}_1(i)E\bar{N}_2(i) \\ &+ \bar{N}_2(i)E'\bar{M}_1(i)B_1(i)R_{11}^{-1}(i)R_{21}(i)R_{11}^{-1}(i)B'_1(i)E = 0, \end{aligned} \tag{5.4.6c}$$

$$A_F(i)E\bar{N}_2(i) + \bar{N}_2(i)E'A'_F(i) + C(i)\bar{N}_2(i)C(i) + \pi_{ii}E'\bar{N}_2(i)E + I_n = 0, \tag{5.4.6d}$$

$$R_{12}(i)F_2(i)\bar{N}_1(i) + R_{22}(i)F_2(i)\bar{N}_2(i) + B'_2(i)(\bar{M}_1(i)E\bar{N}_1(i) + \bar{M}_2(i)E'\bar{N}_2(i)) = 0, \tag{5.4.6e}$$

where

$$F_1(i) = -\sum_{i=1}^{l} R_{11}^{-1}(i)B'_1(i)\bar{M}_1(i)E,\ A_{F_1}(i) = A(i) + B_2(i)F_2(i),\ A_F(i) = A_{F_1}(i) + B_1(i)F_1(i),$$

$$Q_{F_1}(i) = Q_1(i) + F'_2(i)R_{12}(i)F_2(i),\ Q_{F_2}(i) = Q_2(i) + F'_2(i)R_{22}(i)F_2(i).$$

Denote $u_1^*(t) = \sum_{i=1}^{l} F_1(i)\chi_{r_t=i}(t)x(t),\ \ u_2^*(t) = \sum_{i=1}^{l} F_2(i)\chi_{r_t=i}(t)x(t),\ \ i = 1,\ldots,l,$ *then the strategy set* (u_1^*, u_2^*) *constitutes the Stackelberg strategy.*

Proof Given arbitrary $u_2(t) = F_2(r_t)x(t)$, the corresponding u_1 is obtained by minimizing $J_1(x_0, i; u_1)$ with respect to u_1. Let us consider the minimizing problem for the closed-loop stochastic system with arbitrary strategies $u_2(t) = F_2(r_t)x(t)$

$$\begin{aligned}
&\min_{u_1} \bar{J}_1(x_0, i; u_1) = \mathbf{E}\left\{\int_0^\infty \left[x'(t)Q_{F_1}(r_t)x(t) + u_1'(t)R_{11}(r_t)u_1(t)\right]dt \middle| r_0 = i\right\},\\
&s.t.\\
&Edx(t) = [A_{F_2}(r_t)x(t) + B_1(r_t)u_1(t)]dt + C(r_t)x(t)dw(t).
\end{aligned} \tag{5.4.7}$$

By using Theorem 5.1.2, the optimal state feedback controller $u_1^o(t)$ is given by

$$u_1^o(t) = \sum_{i=1}^{l} F_1(i)\chi_{r_t=i}(t)x(t) = -\sum_{i=1}^{l} R_{11}^{-1}(i)B_1'(i)\bar{M}_1(i)\chi_{r_t=i}(t)Ex(t), \tag{5.4.8}$$

where $\bar{M}_1(i)$ is the solution to

$$\begin{aligned}
\mathsf{F}_1(\bar{M}_1(i), F_2(i)) = {} & A_{F_1}'(i)\bar{M}_1(i)E + E'\bar{M}_1(i)A_{F_1}(i) + C'(i)\bar{M}_1(i)C(i)\\
& - F_1'(i)R_{11}(i)F_1(i) + Q_{F_1}(i) + \sum_{j=1}^{l} \pi_{ij}E'\bar{M}_1(j)E = 0.
\end{aligned} \tag{5.4.9}$$

Therefore, Eq. (5.4.6a) holds. On the other hand, if $A_F(i) = A_{F_1}(i) + B_1(i)F_1(i)$ is asymptotically mean square stable, then the cost J_2 of the leader can be represented as

$$J_2(x_0, i; u_1^o(u_2), u_2) = J_2(x_0, i; F_1(r_t)x, F_2(r_t)x(t), = \mathbf{Tr}(\bar{M}_2(i)), \tag{5.4.10}$$

where $\bar{M}_2(i)$ is the solution to

$$\begin{aligned}
\mathsf{F}_2(\bar{M}_1(i), \bar{M}_2(i), F_2(i)) = {} & A_F'(i)\bar{M}_2(i)E + E'\bar{M}_2(i)A_F(i) + C'(i)\bar{M}_2(i)C(i) + Q_{F_2}(i)\\
& + F_1'(i)R_{21}(i)F_1(i) + \sum_{j=1}^{l} \pi_{ij}E'\bar{M}_2(j)E = 0.
\end{aligned} \tag{5.4.11}$$

From (5.4.11) we know (5.4.6b) holds. Let us consider the following Lagrangian **H**

$$\begin{aligned}\mathbf{H}(\bar{M}_1(i),\bar{M}_2(i),F_2(i)) = \mathbf{Tr}(\bar{M}_2(i)) + \mathbf{Tr}(\bar{N}_1(i)\mathsf{F}_1(\bar{M}_1(i),F_2(i))) \\ + \mathbf{Tr}(\bar{N}_2(i)\mathsf{F}_2(\bar{M}_1(i),\bar{M}_2(i),F_2(i))),\end{aligned} \tag{5.4.12}$$

where $\bar{N}_1(i)$ and $\bar{N}_2(i)$ are symmetric matrix of Lagrange multipliers.

As a necessary condition to minimization $\mathbf{Tr}(\bar{M}_2(i))$, we get

$$\begin{aligned}\frac{\partial \mathbf{H}}{\partial \bar{M}_1(i)} &= A_{F_1}(i)\bar{N}_1(i)E + E'\bar{N}_1(i)A'_{F_1}(i) + C(i)\bar{N}_1(i)C'(i) \\ &\quad - E'B_1(i)R_{11}^{-1}(i)B_1'(i)\bar{M}_1(i)E\bar{N}_1(i) - \bar{N}_1(i)E'\bar{M}_1(i)B_1(i)R_{11}^{-1}(i)B_1'(i) \\ &\quad + \pi_{ii}E'\bar{N}_1(i)E - B_1(i)R_{11}^{-1}(i)B_1'(i)\bar{M}_2(i)E\bar{N}_2(i) \\ &\quad - \bar{N}_2(i)E'\bar{M}_2(i)B_1(i)R_{11}^{-1}(i)B_1'(i) \\ &\quad + E'B_1(i)R_{11}^{-1}(i)R_{21}(i)R_{11}^{-1}(i)B_1'(i)\bar{M}_1(i)E\bar{N}_2(i) \\ &\quad + \bar{N}_2(i)E'\bar{M}_1(i)B_1(i)R_{11}^{-1}(i)R_{21}(i)R_{11}^{-1}(i)B_1'(i)E = 0,\end{aligned} \tag{5.4.13a}$$

$$\begin{aligned}\frac{\partial \mathbf{H}}{\partial \bar{M}_2(i)} &= A_F(i)E\bar{N}_2(i) + \bar{N}_2(i)E'A'_F(i) + C(i)\bar{N}_2(i)C(i) \\ &\quad + \pi_{ii}E'\bar{N}_2(i)E + I_n = 0,\end{aligned} \tag{5.4.13b}$$

$$\begin{aligned}\frac{1}{2}\frac{\partial \mathbf{H}}{\partial F_2(i)} &= R_{12}(i)F_2(i)\bar{N}_1(i) + R_{22}(i)F_2(i)\bar{N}_2(i) \\ &\quad + B_2'(i)(\bar{M}_1(i)E\bar{N}_1(i) + \bar{M}_2(i)E'\bar{N}_2(i)) = 0.\end{aligned} \tag{5.4.13c}$$

Therefore, (5.4.6c)–(5.4.6e) hold. This completes the proof of Theorem 5.4.1. □

5.5 Summary

For continuous-time stochastic Markov jump singular systems, we firstly discussed the two person nonzero-sum stochastic differential game problem in finite-time horizon and infinite-time horizon. By using the related conclusion of stochastic LQ problem of Markov jump linear systems, we obtain the necessary and sufficient conditions for the existence of the system combined with Riccati equation method, which corresponds to the existence of the differential (algebraic) Riccati equation, and with the solution of Riccati equation, the optimal control strategy and explicit expression of the optimal value function of the system are given. At the end, two person Stackelberg game problem of stochastic Markov jump singular systems in infinite-time horizon is discussed, and the existence condition of equilibrium strategy and explicit expression are given.

References

1. Wonham, W. M. (1968). On a matrix Riccati equation of stochastic control. *SIAM Journal on Control, 6*(4), 681–697.
2. Athens, M. (1971). Special issues on linear-quadratic-gaussian problem. *IEEE Transactions on Automatic Control, 16*(6), 527–869.
3. Bensoussan, A. (1982). Lectures on stochastic control. In *Nonlinear filtering and stochastic control* (pp. 1–62). Berlin: Springer.
4. Davis, M. H. A. (1977). *Linear estimation and stochastic control*. London: Chapman and Hall.
5. Huang, L., & Mao, X. (2011). Stability of singular stochastic systems with Markovian switching. *IEEE Transactions on Automatic Control, 56*(2), 424–429.
6. Li, X., & Zhou, X. Y. (2002). Indefinite stochastic LQ controls with Markovian jumps in a finite time horizon. *Communications in Information and Systems, 2*(3), 265–282.
7. Zhang, Q., & Xing, S. (2014). Stability analysis and optimal control of stochastic singular systems. *Optimization Letters, 8*(6), 1905–1920.
8. Xu, S., & Lam, J. (2006). *Robust control and filtering of singular systems*. Berlin: Springer.
9. Mukaidani, H. (2014). Stackelberg strategy for discrete-time stochastic system and its application to H_2/H_∞ control. In *American Control Conference (ACC)* (pp. 4488–4493). IEEE.

Chapter 6
Game Theory Approach to Stochastic H_2/H_∞ Control of Markov Jump Singular Systems

In this chapter, we will use the results of stochastic differential games for Markov jump linear systems and Markov jump singular systems to investigate the stochastic H_2/H_∞ robust control problem. First, based on Nash game and Stackelberg game, we studied the stochastic H_2/H_∞ control problem of Markov jump linear systems, the existence conditions for the control strategy and the explicit expression were obtained; then expand to the corresponding results to Markov jump singular systems, and based on Nash game and Stackelberg game, the existence conditions for the optimal control strategy of Markov jump singular systems and the explicit expression were given.

6.1 Introduction

As the old saying, "Anything unexpected may happen, people have always happens", the real world is full of uncertainty. To cope with the possible impact of uncertainty, people invented various coping methods, and robust control is an effective method for processing uncertainty.

In modern robust control theory, H_2/H_∞ control problem caused widespread concern of scholars, and has been widely used in various fields. Game theory approach is an important method among all the methods in dealing with H_2/H_∞ control problem. The basic idea of the robust control based on game theory is that: the designer of the control strategy that is regarded as one of the player P_1, the other uncertain or disturbance that is regarded as another one of the player "natural" P_2, so that the H_2/H_∞ robust control problem can be converted into a two person game problem, P_1 faced the problem that how to design his own strategies in various anticipated disturbances to balance with "natural" P_2 and make his own goals best. Then, using the Nash equilibrium strategy or Stackelberg strategy to obtain the optimal control strategy.

C.-k. Zhang et al., *Non-cooperative Stochastic Differential Game Theory of Generalized Markov Jump Linear Systems*, Studies in Systems, Decision and Control 67, DOI 10.1007/978-3-319-40587-2_6

Inspired by the above method, this chapter discuss the stochastic H_2/H_∞ control problem of Markov jump linear systems and Markov jump singular systems based on Nash game and Stackelberg game approach, the existence conditions of the equilibrium strategy and design methods are given, and the explicit expression of the equilibrium strategy is also obtained.

6.2 Stochastic H_2/H_∞ Control to Markov Jump Linear System Based on Nash Games

6.2.1 *Finite-Time Horizon Case*

Given a filtered probability space $(\Omega, \mathcal{F}, \{\mathcal{F}_t\}_{t\geq 0}, \mathcal{P})$, on which there exists a one-dimensional standard $\mathcal{F}_t$–Brownian motion $w(t)$ on $[0, T]$ (with $w(0) = 0$), and a Markov chain r_t which is adapted to $\mathcal{F}_t$, taking values in $\Xi = \{1, \ldots, l\}$, with the transition probabilities given by:

$$\Pr\{r_{t+\Delta} = j | r_t = i\} = \begin{cases} \pi_{ij}\Delta + o(\Delta), & \text{if } i \neq j, \\ 1 + \pi_{ij}\Delta + o(\Delta), & \text{else}, \end{cases} \tag{6.2.1}$$

where $\pi_{ij} \geq 0$ for $i \neq j$ and $\pi_{ii} = -\sum_{i\neq j} \pi_{ij}$. In addition, we assume that the processes r_t and $w(t)$ are independent.

Let $L^2_{\mathcal{F}}(0, T; \mathbb{R}^n) := \{\phi(\cdot,\cdot) : [0, T] \times \Omega \to \mathbb{R}^n | \phi(\cdot,\cdot)$ is an $\mathcal{F}_t$—adapted process on $[0, T]$, and $\mathbf{E}\int_0^T \|\phi(t,\omega)\|^2 dt < \infty\}$.

Consider the following continuous-time stochastic Markov jump systems with state-, control- and disturbance-dependent noise

$$\begin{cases} dx(t) = [A_1(t, r_t)x(t) + B_1(t, r_t)u(t) + C_1(t, r_t)v(t)]dt \\ \qquad\qquad + [A_2(t, r_t)x(t) + B_2(t, r_t)u(t) + C_2(t, r_t)v(t)]dw(t), \\ z(t) = \begin{bmatrix} D(t, r_t)x(t) \\ F(t, r_t)u(t) \end{bmatrix}, \\ F'(t, r_t)F(t, r_t) = I, x(0) = x_0 \in \mathbb{R}^n, t \in [0, T]. \end{cases} \tag{6.2.2}$$

where $x(t) \in \mathbb{R}^n$, $u(t) \in \mathbb{R}^{n_u}$, $v(t) \in \mathbb{R}^{n_v}$, and $z(t) \in \mathbb{R}^{n_z}$ are the system state, control input, exogenous input, and regulated output, respectively. All coefficients of (6.2.2) are assumed to be continuous matrix-valued functions of suitable dimensions.

To give our main results in the next subsection, we need the following definitions and lemmas. Given disturbance attenuation $\gamma > 0$, define two associated performances as follows, $i \in \Xi$:

$$J_1(u,v;x_0,r_0) = \gamma^2\|v(t)\|^2_{[0,T]} - \|z(t)\|^2_{[0,T]} = \mathbf{E}\left\{\int_0^T \left(\gamma^2\|v(t)\|^2 - \|z(t)\|^2\right)dt | r_0 = i\right\}, \tag{6.2.3a}$$

$$J_2(u,v;x_0,r_0) = \|z(t)\|^2_{[0,T]} = \mathbf{E}\left\{\int_0^T \|z(t)\|^2 dt | r_0 = i\right\}. \tag{6.2.3b}$$

The definition of finite-time horizon stochastic H_2/H_∞ control problem is:

Definition 6.2.1 For system (6.2.1) and given $\gamma > 0$, $0 < T < \infty$, find, if possible, a state feedback control $u^*(t) \in L^2_{\mathcal{F}}(0,T;\mathbb{R}^{n_u})$ such that:

(i)
$$\|\mathcal{L}_T\| = \sup_{\substack{v \neq 0 \\ u = u^* \\ x_0 = 0}} \frac{\mathbf{E}\left\{\int_0^T \left(\|D(t,r_t)x(t)\|^2 + \|u^*(t)\|^2\right)dt | r_0 = i\right\}^{\frac{1}{2}}}{\mathbf{E}\left\{\int_0^T \|v(t)\|^2 dt | r_0 = i\right\}^{\frac{1}{2}}} < \gamma, \tag{6.2.4}$$

where $i \in \Xi$, and $\mathcal{L}_T$ is an operator associated with system (6.2.1) which is defined as

$$\mathcal{L}_T : L^2_{\mathcal{F}}(0,T;\mathbb{R}^{n_v}) \mapsto L^2_{\mathcal{F}}(0,T;\mathbb{R}^{n_z}),$$
$$\mathcal{L}_T(v(t)) = z(t)|_{x_0=0}, t \in [0,T].$$

(ii) When the worst case disturbance $v^*(t) \in L^2_{\mathcal{F}}(0,T;\mathbb{R}^{n_v})$, if it exists, is applied to (6.2.1), $u^*(t)$ minimizes the output energy

$$J_2(u,v^*;x_0,r_0) = \mathbf{E}\left\{\int_0^T \left(\|D(t,r_t)x(t)\|^2 + \|u(t)\|^2\right)dt | r_0 = i\right\}, \tag{6.2.5}$$

where $v^*(t)$ is defined as

$$v^*(t) = \arg\min\left\{J_1(u^*,v;x_0,r_0) = \mathbf{E}\left\{\int_0^T \left(\gamma^2\|v(t)\|^2 - \|z(t)\|^2\right)dt | r_0 = i\right\}\right\}.$$

If the above (u^*, v^*) exist, then we say that the finite-time horizon H_2/H_∞ control of system (6.2.1) is solvable and has a pair of solutions (u^*, v^*).

In other words, for given two cost functions defined in (6.2.3a) and (6.2.3b), the finite-time horizon H_2/H_∞ control of system (6.2.1) is equivalent to finding the Nash equilibrium (u^*, v^*), such that

$$\begin{aligned} &J_1(u^*,v^*;x_0,r_0)\le J_1(u^*,v;x_0,r_0),\ J_2(u^*,v^*;x_0,r_0)\le J_2(u,v^*;x_0,r_0),\\ &\quad (u(t),v(t))\in L^2_{\mathcal{F}}(0,T;\mathbb{R}^{n_u})\times L^2_{\mathcal{F}}(0,T;\mathbb{R}^{n_v}),\ r_0\in\Xi. \end{aligned} \tag{6.2.6}$$

The first inequality of (6.2.6) is associated with the H$_\infty$ performance, while the second one is related with the H$_2$ performance. Clearly, if the Nash equilibrium (u^*,v^*) exist, u^* is our desired H$_2$/H$_\infty$ controller, and v^* is the worst case disturbance. In this case, we also say that the stochastic H$_2$/H$_\infty$ control admits a pair of solutions (u^*,v^*).

Before giving the main results, some preliminary work needs to be introduced. Consider the following stochastic perturbed system with Markov jump parameters

$$\begin{cases} dx(t)=[A_1(t,r_t)x(t)+C_1(t,r_t)v(t)]dt+[A_2(t,r_t)x(t)+C_2(t,r_t)v(t)]dw(t),\\ z(t)=D(t,r_t)x(t),\quad x(0)=x_0\in\mathbb{R}^n,\quad t\in[0,T]. \end{cases} \tag{6.2.7}$$

For any given $0<T<\infty$, associated with system (6.2.7), the perturbation operator $\tilde{\mathcal{L}}_T: L^2_{\mathcal{F}}(0,T;\mathbb{R}^{n_v})\mapsto L^2_{\mathcal{F}}(0,T;\mathbb{R}^{n_z})$ is defined as $\tilde{\mathcal{L}}_T(v(t))=z(t)|_{x_0=0}=D(t,r_t)x(t)|_{x_0=0}$, $t\in[0,T]$, then

$$\|\tilde{\mathcal{L}}_T\|=\sup_{v\neq 0,x_0=0}\frac{\mathbf{E}\left\{\int_0^T\|D(t,r_t)x(t)\|^2dt|r_0=i\right\}^{1/2}}{\mathbf{E}\left\{\int_0^T\|v(t)\|^2dt|r_0=i\right\}^{1/2}}.$$

In our subsequent analysis, we define $M_i=M(t,i)$, $M_{1i}=M_1(t,i)$, $M_{2i}=M_2(t,i)$, $i\in\Xi$ for convenience.

Lemma 6.2.1 *For system (6.2.7) and given disturbance attenuation $\gamma>0$, $\|\tilde{\mathcal{L}}_T\|<\gamma$ iff the following coupled generalized differential Riccati equations*

$$\begin{cases} \dot{P}_i+P_iA_{1i}+A_{1i}'P_i+A_{2i}'P_iA_{2i}-D_i'D_i+\sum\limits_{j=1}^{l}\pi_{ij}P_j\\ -(P_iC_{1i}+A_{2i}'P_iC_{2i})\left(\gamma^2I+C_{2i}'P_iC_{2i}\right)^{-1}(C_{1i}'P_i+C_{2i}'P_iA_{2i})=0,\\ P(T,i)=0,\\ \gamma^2I+C_{2i}'P_iC_{2i}>0,\ \forall t\in[0,T],\quad i\in\Xi \end{cases} \tag{6.2.8}$$

have a bounded solution $P(t)=(P(t,1),\ldots,P(t,l))\le 0\in\mathcal{C}([0,T];\mathcal{S}_l^n)$.

Proof The details are similar with Ref. [1], so we omitted it here.

The following theorem presents the main results of finite-time horizon stochastic H$_2$/H$_\infty$ control. □

Theorem 6.2.1 *For system (6.2.1), given a disturbance attenuation level $\gamma>0$ and $0<T<\infty$, the stochastic H$_2$/H$_\infty$ control admits a pair of solutions (u^*,v^*) with*

$$u^*(t) = \sum_{i=1}^{l} K_2(t,i)\chi_{r_t=i}(t)x(t), v^*(t) = \sum_{i=1}^{l} K_1(t,i)\chi_{r_t=i}(t)x(t), \tag{6.2.9}$$

if and only if for $\forall t \in [0,T]$, $i \in \Xi$, *the following four coupled differential Riccati equations*

$$\begin{cases} \dot{P}_i^1 + P_i^1\tilde{A}_{1i} + \tilde{A}_{1i}'P_i^1 + \tilde{A}_{2i}'P_i^1\tilde{A}_{2i} + \tilde{Q}(i) + \sum_{j=1}^{l}\pi_{ij}P_j^1 \\ -(P_i^1C_{1i} + \tilde{A}_{2i}'P_i^1C_{2i})(\gamma^2 I + C_{2i}'P_i^1C_{2i})^{-1}(C_{1i}'P_i^1 + C_{2i}'P_i^1\tilde{A}_{2i}) = 0, \\ P^1(T,i) = 0, \\ \gamma^2 I + C_{2i}'P_i^1C_{2i} > 0, i \in \Xi. \end{cases} \tag{6.2.10}$$

$$K_{1i} = -(\gamma^2 I + C_{2i}'P_i^1C_{2i})^{-1}(C_{1i}'P_i^1 + C_{2i}'P_i^1\tilde{A}_{2i}). \tag{6.2.11}$$

$$\begin{cases} \dot{P}_i^2 + P_i^2\bar{A}_{1i} + \bar{A}_{1i}'P_i^2 + \bar{A}_{2i}'P_i^2\bar{A}_{2i} + D_i'D_i + \sum_{j=1}^{l}\pi_{ij}P_j^2 \\ -(P_i^2B_{1i} + \bar{A}_{2i}'P_i^2B_{2i})(I + B_{2i}'P_i^2B_{2i})^{-1}(B_{1i}'P_i^2 + B_{2i}'P_i^2\bar{A}_{2i}) = 0, \\ P^2(T,i) = 0, \\ I + B_{2i}'P_i^2B_{2i} > 0, \ i \in \Xi. \end{cases} \tag{6.2.12}$$

$$K_{2i} = -(I + B_{2i}'P_i^2B_{2i})^{-1}(B_{1i}'P_i^2 + B_{2i}'P_i^2\bar{A}_{2i}), \tag{6.2.13}$$

where

$$\begin{aligned} \tilde{A}_{1i} &= A_{1i} + B_{1i}K_{2i}, \tilde{A}_{2i} = A_{2i} + B_{2i}K_{2i}, \tilde{Q}_i = -(D_i'D_i + K_{2i}'K_{2i}), \\ \bar{A}_{1i} &= A_{1i} + C_{1i}K_{1i}, \bar{A}_{2i} = A_{2i} + C_{2i}K_{1i} \end{aligned}$$

admit a solution set $(P^1(t), P^2(t), K_1(t), K_2(t))$, *with*

$$\begin{aligned} P^1(t) &= (P^1(t,1), \ldots, P^1(t,l)) \le 0 \in \mathcal{C}([0,T]; \mathcal{S}_l^n), \\ P^2(t) &= (P^2(t,1), \ldots, P^2(t,l)) \ge 0 \in \mathcal{C}([0,T]; \mathcal{S}_l^n) \end{aligned}$$

Proof Sufficiency: $u^*(t) = \sum_{i=1}^{l} K_2(t,i)\chi_{r_t=i}(t)x(t)$ into (6.2.2), it follows that

$$\begin{cases} dx(t) = [\tilde{A}_1(t,r_t)x(t) + C_1(t,r_t)v(t)]dt + [\tilde{A}_2(t,r_t)x(t) + C_2(t,r_t)v(t)]dw(t), \\ z(t) = \begin{bmatrix} D(t,r_t)x(t) \\ K_2(t,r_t)x(t) \end{bmatrix}, x(0) = x_0 \in \mathbb{R}^n. \end{cases} \tag{6.2.14}$$

Considering Eq. (6.2.10), using Lemma 6.2.1 to system (6.2.14) immediately yields $\|\mathcal{L}_T^{K_2}\| < \gamma$. And from

$$\begin{aligned}
J_1(u^*, v; x_0, r_0) &= \mathbf{E}\left\{\int_0^T \left(\gamma^2\|v(t)\|^2 - \|z(t)\|^2\right) dt | r_0 = i\right\} \\
&= x_0' P^1(0, r_0) x_0 + \mathbf{E}\left\{\int_0^T \left[\gamma^2\|v(t)\|^2 - \|z(t)\|^2 + d(x'(t) P^1(t, r_t) x(t)) dt | r_0 = i\right]\right\} \\
&= x_0' P^1(0, r_0) x_0 + \mathbf{E}\left\{\int_0^T (v(t) - v^*(t))' (\gamma^2 I + C_2'(t, r_t) P^1(t, r_t) C_2(t, r_t)) (v(t) - v^*(t)) dt | r_0 = i\right\} \\
&\geq J_1(u^*, v^*; x_0, r_0) = x_0' P^1(0, r_0) x_0,
\end{aligned}$$

we can see that $v^*(t) = \sum_{i=1}^{l} K_1(i) \chi_{r_t=i}(t) x(t)$ is the worst case disturbance.

Now, substituting $v = v^*(t) = \sum_{i=1}^{l} K_1(i) \chi_{r_t=i}(t) x(t)$ into (6.2.2), it follows that

$$\begin{cases} dx(t) = [\bar{A}_1(t, r_t)x + B_1(t, r_t)u(t)]dt + [\bar{A}_2(t, r_t)x + B_2(t, r_t)u(t)]dw(t), \\ z(t) = \begin{bmatrix} D(t, r_t)x(t) \\ u(t) \end{bmatrix}, x(0) = x_0 \in \mathbb{R}^n. \end{cases} \tag{6.2.15}$$

With the constraint of (6.2.15), minimizing $J_2(u, v^*; x_0, r_0)$ is a standard stochastic linear quadratic optimization problem. Applying a standard completion of square technique together with considering (6.2.12), we have

$$\min_{u \in L^2_{\mathcal{F}}(0,T;\mathbb{R}^{n_u})} J_2(u, v^*; x_0, r_0) = J_2(u^*, v^*; x_0, r_0) = x_0' P^2(0, r_0) x_0$$

with the corresponding optimal control $u^*(t) = \sum_{i=1}^{l} K_2(t, i) \chi_{r_t=i}(t) x(t)$. The sufficiency is proved.

Necessity: If $u^*(t) = \sum_{i=1}^{l} K_2(t, i) \chi_{r_t=i}(t) x(t)$ and $v^*(t) = \sum_{i=1}^{l} K_1(i) \chi_{r_t=i}(t) x(t)$ solves the finite-time horizon stochastic H$_2$/H$_\infty$ control, where K_1 and K_2 are to be determined, then substituting $u^*(t) = \sum_{i=1}^{l} K_2(t, i) \chi_{r_t=i}(t) x(t)$ into (6.2.2) results in (6.2.14). Lemma 6.2.1 concludes that (6.2.10) admits a solution $P^1(t) = (P^1(t, 1), \ldots, P^1(t, l)) \leq 0 \in \mathcal{C}([0, T]; \mathcal{S}_l^n)$, with $v^*(t) = \sum_{i=1}^{l} K_1(i) \chi_{r_t=i}(t) x(t)$, where K_1 is defined by (6.2.11). Likewise, if we implement v_T^* in (6.2.2), it deduces (6.2.15). While Eq. (6.2.12) always exists a solution $P^2(t) = (P^2(t, 1), \ldots, P^2(t, l)) \geq 0 \in \mathcal{C}([0, T]; \mathcal{S}_l^n)$ for fixed K_1, see Ref. [2]. As discussion in the sufficiency part, in this case, $u^*(t) = \sum_{i=1}^{l} K_2(t, i) \chi_{r_t=i}(t) x(t)$ is optimal, with K_2 defined by (6.2.13). The Necessity is proved.

So this ends the proof of Theorem 6.2.1. □

6.2.2 Infinite-Time Horizon Case

6.2.2.1 Preliminaries

Consider the following controlled linear stochastic system with Markovian jumps

$$dx(t) = [A(r_t)x(t) + B(r_t)u(t)]dt + A_1(r_t)x(t)dw(t) \tag{6.2.16}$$

where $x(t) \in \mathbb{R}^n$ is the system state, $u(t) \in \mathbb{R}^{n_u}$ is control input, all coefficient matrices are assumed to be constant with compatible dimensions for given $r_t = i \in \Xi$.

For system (6.2.16), applying generalized Itô's formula to $x'P(i)x$, we have

Lemma 6.2.2 *Suppose $P = (P(1), P(2), \ldots, P(l)) \in \mathcal{S}_l^n$ is given, then for system (6.2.16) with initial state $(x_0, i) \in \mathbb{R}^n \times \Xi$, we have for any $T > 0$,*

$$\mathbf{E}\left\{\int_0^T [x'(t)(P(r_t)A(r_t) + A'(r_t)P(r_t) + A_1'(r_t)P(r_t)A_1(r_t) + \sum_{j=1}^{l} \pi_{r_t j}P(j))x(t) \right.$$
$$\left. + 2u'(t)B'(r_t)P(r_t)x(t)]dt | r_0 = i\right\} = \mathbf{E}[x'(T)P(r_T)x(T)|r_0 = i] - x_0'P(i)x_0. \tag{6.2.17}$$

6.2.2.2 Main Results

For simplicity and without loss of generality, consider the following Markov jump systems described by stochastic differential equation

$$\begin{cases} dx(t) = [A(r_t)x(t) + B_2(r_t)u(t) + B_1(r_t)v(t)]dt + [A_1(r_t)x(t) + C_2(r_t)u(t)]dw(t), \\ z(t) = \begin{bmatrix} C_0(r_t)x(t) \\ D(r_t)u(t) \end{bmatrix}, D'(i)D(i) \equiv I, x(0) = x_0 \in \mathbb{R}^n, i \in \Xi. \end{cases} \tag{6.2.18}$$

where $x(t) \in \mathbb{R}^n$, $u(t) \in \mathbb{R}^{n_u}$, $v(t) \in \mathbb{R}^{n_v}$, and $z(t) \in \mathbb{R}^{n_z}$ are the system state, control input, exogenous input, and regulated output, respectively. All coefficients of (6.2.18) are assumed to be constants. Define two associated cost functions as follows:

$$J_1^\infty(u, v; x_0, i) = \mathbf{E}\left\{\int_0^\infty \left[\gamma^2\|v(t)\|^2 - \|z(t)\|^2\right] dt | r_0 = i\right\} \tag{6.2.19a}$$

$$J_2^\infty(u, v; x_0, i) = \mathbf{E}\left\{\int_0^\infty \|z(t)\|^2 dt | r_0 = i\right\}, \quad i \in \Xi \tag{6.2.19b}$$

The infinite-time horizon stochastic H$_2$/H$_\infty$ control problem of system (6.2.18) is described as follows.

Definition 6.2.4 [3] For given disturbance attenuation level $\gamma > 0$, if we can find $u^*(t) \times v^*(t) \in L^2_{\mathcal{F}}(0,\infty;\mathbb{R}^{n_u}) \times L^2_{\mathcal{F}}(0,\infty;\mathbb{R}^{n_v})$, such that

(i) $u^*(t)$ stabilizes system (6.2.18) internally, i.e. when $v(t) = 0$, $u = u^*$, the state trajectory of (6.2.18) with any initial value $(x_0, i) \in \mathbb{R}^n \times \Xi$ satisfies

$$\lim_{t\to\infty} \mathbf{E}\left[\|x(t)\|^2 | r_0 = i\right] = 0.$$

(ii) $|\mathcal{L}_{u^*}|_\infty < \gamma$ with

$$\|\mathcal{L}_{u^*}\|_\infty = \sup_{\substack{v \in L^2_{\mathcal{F}}(0,\infty;\mathbb{R}^{n_v}) \\ v \neq 0, u = u*, x_0 = 0}} \frac{\mathbf{E}\left\{\int_0^\infty \left[\|C_0(r_t)x(t)\|^2 + \|u^*(t)\|^2\right] | r_0 = i\right\}^{1/2}}{\mathbf{E}\left\{\int_0^\infty \|v(t)\|^2 dt | r_0 = i\right\}^{1/2}}.$$

(iii) When the worst case disturbance $v^*(t) \in L^2_{\mathcal{F}}(0,\infty;\mathbb{R}^{n_v})$, if existing, is applied to (6.2.18), $u^*(t)$ minimizes the output energy

$$J_2^\infty(u, v^*; x_0, i) = \mathbf{E}\left\{\int_0^\infty \left[\|C_0(r_t)x(t)\|^2 + \|u(t)\|^2\right] dt | r_0 = i\right\}, \quad i \in \Xi,$$

where $v^*(t)$ is defined as

$$v^*(t) = \arg\min\left\{J_1^\infty(u^*, v; x_0, i) = \mathbf{E}\left\{\int_0^\infty \left(\gamma^2\|v(t)\|^2 - \|z(t)\|^2\right) dt | r_0 = i\right\}\right\}.$$

If the above (u^*, v^*) exist, then we say that the infinite-time horizon H$_2$/H$_\infty$ control of system (6.2.18) is solvable and has a pair of solutions. Obviously, (u^*, v^*) is the Nash equilibrium strategies such that

$$J_1^\infty(u^*, v^*; x_0, i) \leq J_1^\infty(u^*, v; x_0, i), \tag{6.2.20}$$

$$J_2^\infty(u^*, v^*; x_0, i) \leq J_2^\infty(u, v^*; x_0, i), \quad i \in \Xi. \tag{6.2.21}$$

The main result of the infinite-time horizon stochastic H$_2$/H$_\infty$ control is presented by the following theorem, which can be shown following the line of Theorem 2.1 and Theorem 1 presented in Ref. [4].

Theorem 6.2.2 *For a given disturbance attenuation $\gamma > 0$, suppose systems (6.2.18) is stochastically stablizable, infinite-time horizon stochastic H$_2$/H$_\infty$ control has a pair of solutions (u^*, v^*) with $u^*(t) = K_2(r_t)x(t)$ and $v^*(t) = K_1(r_t)x(t)$, where $K_2(i) \in$*

$\mathcal{M}^l_{n_u,n}$ and $K_1(i) \in \mathcal{M}^l_{n_v,n}$ are constant matrices, if and only if the following four coupled algebraic Riccati equations (6.2.22) *admit solutions* (P_1, P_2, K_1, K_2) *as* $P_1 = (P_1(1), P_1(2), \ldots, P_1(l)) \le 0 \in \mathcal{S}^n_l, P_2 = (P_2(1), P_2(2), \ldots, P_2(l)) \ge 0 \in \mathcal{S}^n_l$.

$$\begin{aligned} &P_1(i)\tilde{A}(i) + \tilde{A}'(i)P_1(i) + \tilde{A}_1'(i)P_1(i)\tilde{A}_1(i) + \tilde{Q}(i) + \sum_{j=1}^{l} \pi_{ij}P_1(j) \\ &- \gamma^{-2}P_1(i)B_1(i)B_1'(i)P_1(i) = 0,\ i = 1, \ldots, l. \end{aligned} \tag{6.2.22a}$$

$$K_1(i) = -\gamma^{-2}B_1'(i)P_1(i). \tag{6.2.22b}$$

$$\begin{cases} P_2(j)\bar{A}(j) + \bar{A}'(j)P_2(j) + A_1'(j)P_2(j)A_1(j) + C_0'(j)C_0(j) + \sum\limits_{k=1}^{l} \pi_{jk}P_2(k) \\ -\left(P_2(j)B_2(j) + A_1'(j)P_2(j)C_2(j)\right)\left(I + C_2'(j)P_2(j)C_2(j)\right)^{-1} \\ \times\left(B_2'(j)P_2(j) + C_2'(j)P_2(j)A_1(j)\right) = 0, \\ I + C_2'(j)P_2(j)C_2(j) > 0, j \in \Xi. \end{cases} \tag{6.2.22c}$$

$$K_2(j) = -\left(I + C_2'(j)P_2(j)C_2(j)\right)^{-1}\left(B_2'(j)P_2(j) + C_2'(j)P_2(j)A_1(j)\right), \tag{6.2.22d}$$

where

$$\tilde{A} = A - B_2K_2, \tilde{A}_1 = A_1 - C_2K_2, \tilde{Q} = -\left(C_0'C_0 + K_2'K_2\right), \bar{A} = A - B_1K_1.$$

6.2.3 Numerical Examples

In order to verify the correctness of the conclusions, consider all the coefficient matrices of the system (6.2.18) taking the following values:

$$\Xi = \{1,2\}, \Pi = \begin{bmatrix} -0.2 & 0.2 \\ 0.8 & -0.8 \end{bmatrix}, A(1) = \begin{bmatrix} 0 & 1 \\ -2 & -3 \end{bmatrix}, A(2) = \begin{bmatrix} 0 & 1 \\ 1 & 0 \end{bmatrix}, B_1(1) = \begin{bmatrix} 1 \\ 1 \end{bmatrix},$$
$$B_1(2) = \begin{bmatrix} 0 \\ 1 \end{bmatrix}, B_2(1) = \begin{bmatrix} 1 \\ 0 \end{bmatrix}, B_2(2) = \begin{bmatrix} 3 \\ 0 \end{bmatrix}, A_1(1) = \begin{bmatrix} 0.1 & 0 \\ 0 & 0.3 \end{bmatrix}, A_1(2) = \begin{bmatrix} 0.5 & 0 \\ 0 & 0.2 \end{bmatrix},$$
$$C_2(1) = \begin{bmatrix} 0 \\ 0.1 \end{bmatrix}, C_2(2) = \begin{bmatrix} 0 \\ 0 \end{bmatrix}.$$

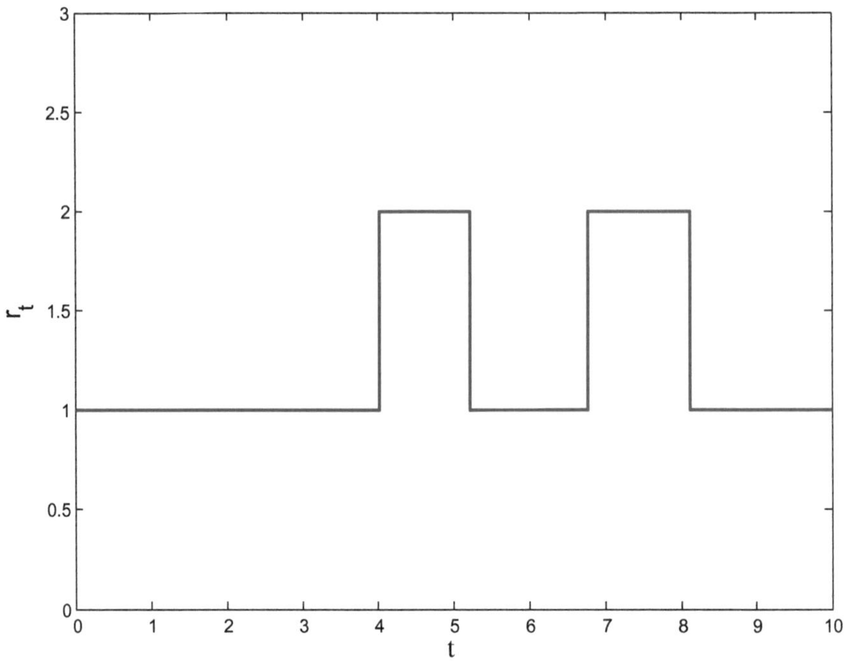

Fig. 6.1 Curve of r_t

setting $\gamma = 0.7$, applying the algorithm proposed in Ref. [5] to (6.2.22a)–(6.2.22d), we have

$$P(1) = \begin{bmatrix} 0.0348 & 0.0246 \\ 0.0246 & 0.0512 \end{bmatrix}, P(2) = \begin{bmatrix} 0.0427 & 0.0682 \\ 0.0682 & 0.3112 \end{bmatrix}.$$

So, the stochastic H_2/H_∞ control strategy is:

$$u(t) = -0.0350x_1(t) - 0.0261x_2(t), \text{when } r_t = 1;$$
$$u(t) = -0.1281x_1(t) - 0.2046x_2(t), \text{when } r_t = 2.$$

Using Matlab with simulation step $\Delta = 0.001$, initial value $r_0 = 1$, $x_1(0) = 2$ and $x_2(0) = 1$, we obtain the state trajectories as shown in Figs. 6.1, 6.2 and 6.3:

As can be seen from Figs. 6.1, 6.2 and 6.3, under the control of $u(t)$, the closed-loop system is stable.

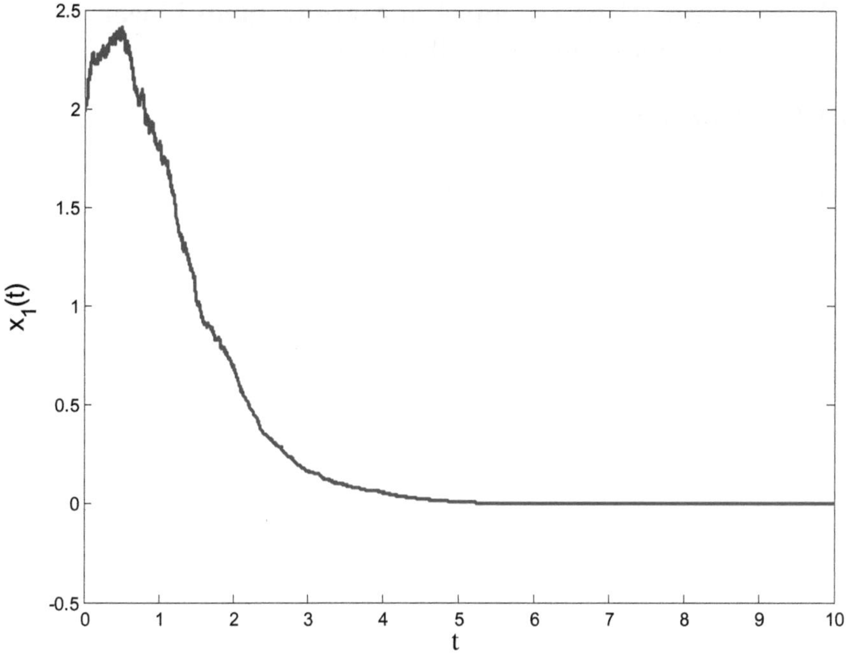

Fig. 6.2 Curve of $x_1(t)$

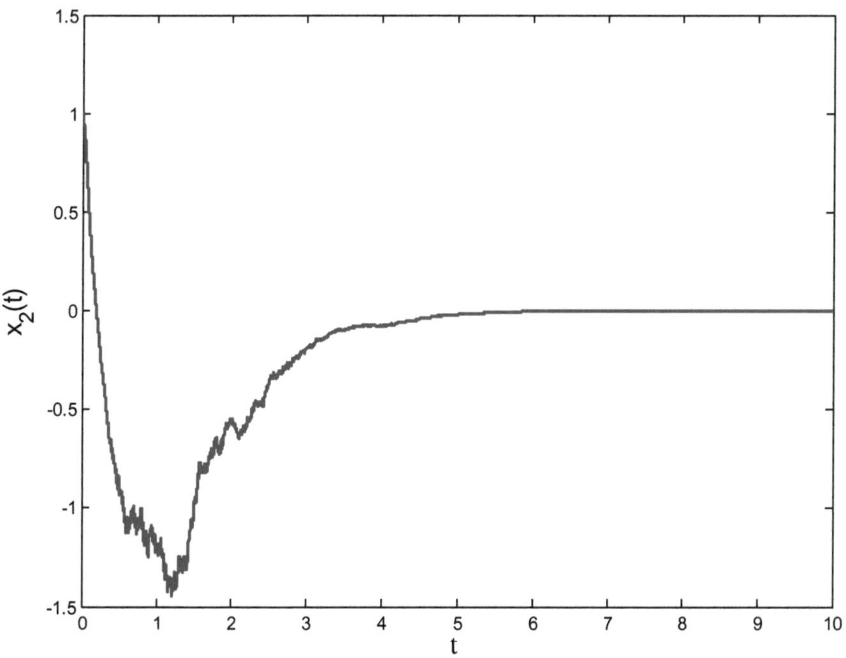

Fig. 6.3 Curve of $x_2(t)$

6.3 Stochastic H2/H∞ Control to Markov Jump Linear Systems Based on Stackelberg Game

6.3.1 Preliminary Results

Consider the following Markov jump linear systems

$$\begin{cases} dx(t) = [A(r_t)x(t) + D(r_t)v(t)]dt + A_p(r_t)x(t)dw(t), \\ z(t) = C(r_t)x(t), \quad x(0) = x_0 \in \mathbb{R}^n, \end{cases} \tag{6.3.1}$$

where $v(t) \in \mathbb{R}^{n_v}$ and $z(t) \in L^2_{\mathcal{F}}(0, \infty; \mathbb{R}^{n_z})$ are, respectively, disturbance signal and controlled output. For system (6.3.1), the perturbed operator $\mathcal{L} : L^2_{\mathcal{F}}(0, \infty; \mathbb{R}^{n_v}) \mapsto L^2_{\mathcal{F}}(0, \infty; \mathbb{R}^{n_z})$ of $\mathcal{L}(v(t)) = z(t)|_{x_0=0} = C(r_t)x(t)|_{x_0=0}$ is defined by

$$\|\mathcal{L}\|_\infty = \sup_{v \neq 0, x_0 = 0} \frac{\mathbf{E}\left\{\int_0^\infty \|z(t)\|^2 dt | r_0 = i\right\}^{1/2}}{\mathbf{E}\left\{\int_0^\infty \|v(t)\|^2 dt | r_0 = i\right\}^{1/2}}.$$

The following real bounded lemma was proposed by Huang and Zhang [3]:

Lemma 6.3.1 *Given a disturbance attenuation level $\gamma > 0$, the stochastic system (6.3.1) is internal stable and $\|\mathcal{L}\|_\infty < \gamma$ for some γ, if and only if (iff) the following equation*

$$\begin{aligned} &A'(i)P(i) + P(i)A(i) + A_p'(i)P(i)A_p(i) + C'(i)C(i) \\ &+ \gamma^{-2}P(i)D(i)D'(i)P(i) + \sum_{j=1}^{l} \pi_{ij}P(j) = 0 \end{aligned} \tag{6.3.2}$$

has a solution $P = (P(1), \ldots, P(l)) > 0 \in \mathcal{S}_l^n$.

In this case, $v^*(t) = \sum_{i=1}^{l} F^*(i)\chi_{r_t=i}(t)x(t) = \gamma^{-2}\sum_{i=1}^{l} D'(i)P(i)\chi_{r_t=i}(t)x(t)$ is the worst disturbance of the system.

6.3.2 Problem Formulation

Consider the following controlled Markov jump linear systems

$$\begin{cases} dx(t) = [A(r_t)x(t) + B(r_t)u(t) + D(r_t)v(t)]dt + A_p(r_t)x(t)dw(t), \\ z(t) = C(r_t)x(t), x(0) = x_0. \end{cases} \tag{6.3.3}$$

where $x(t) \in \mathbb{R}^n$, $u(t) \in \mathbb{R}^{n_u}$, $v(t) \in \mathbb{R}^{n_v}$ and $z(t) \in \mathbb{R}^{n_z}$ are, respectively, the system state, control input, disturbance signal and controlled output. In (6.3.3), $A(r_t) = A(i)$, $B(r_t) = B(i)$, $C(r_t) = C(i)$, $D(r_t) = D(i)$ and $A_p(r_t) = A_p(i)$, when $r_t = i$, while $A(i)$, etc., $i = 1, \cdots, l$, are constant matrices with suitable size.

The definition of stochastic H$_2$/H$_\infty$ control based on Stackelberg game is given below:

Definition 6.3.1 [5] If there exists a strategy set (v^*, u^*), such that

$$J_2(x_0, i; v^*, u^*) \leq J_2(x_0, i; v^o(u), u), \quad \forall u \in \mathbb{R}^{n_u}, \tag{6.3.4}$$

where

$$J_1(x_0, i; v^o(u), u) = \min_v J_1(x_0, i; v, u), \tag{6.3.5}$$

and

$$v^* = v^{\circ}(u^*) \tag{6.3.6a}$$

$$J_1(x_0, i; v, u) = \mathbf{E}\left\{ \int_0^\infty \left[\gamma^2 \|v(t)\|^2 - \|z(t)\|^2 \right] dt | r_0 = i \right\} \tag{6.3.6b}$$

$$\begin{aligned} J_2(x_0, i; v, u) &= \mathbf{E}\left\{ \int_0^\infty [x'(t)Q(r_t)x(t) + u'(t)R(r_t)u(t)] dt | r_0 = i \right\}, \\ Q(r_t) &= Q'(r_t) \geq 0, \quad R(r_t) = R'(r_t) > 0, \end{aligned} \tag{6.3.6c}$$

then, this strategy set is the desired stochastic H$_2$/H$_\infty$ control set.

Without loss of generality, we restrict our strategies to linear state feedback case, i.e., the closed-loop stochastic H$_2$/H$_\infty$ sets have the following form

$$v(t) = \sum_{i=1}^{l} F_\gamma(i) \chi_{r_t=i}(t) x(t), u(t) = \sum_{i=1}^{l} K(i) \chi_{r_t=i}(t) x(t).$$

6.3.3 Main Results

The following theorem presents the main results of stochastic H$_2$/H$_\infty$ control based on Stackelberg strategy:

Theorem 6.3.1 Suppose that the following cross-coupled matrix-valued equations (6.3.7a)–(6.3.7e) has solutions $\bar{M}_1(i) \leq 0$, $\bar{M}_2(i) \geq 0$, $\bar{N}_j(i), j = 0, 1$, $F_\gamma(i)$ and $K(i)$, $i = 1, \cdots l$

$$A_K'(i)\bar{M}_1(i) + \bar{M}_1(i)A_K(i) + A_p'(i)\bar{M}_1(i)A_p(i) \\ - \gamma^{-2}\bar{M}_1(i)D'(i)D(i)\bar{M}_1(i) - C'(i)C(i) + \sum_{j=1}^{l}\pi_{ij}\bar{M}_1(j) = 0, \tag{6.3.7a}$$

$$A_F'(i)\bar{M}_2(i) + \bar{M}_2(i)A_F(i) + A_p'(i)\bar{M}_2(i)A_p(i) + Q(i) \\ + K'(i)R(i)K(i) + \sum_{j=1}^{l}\pi_{ij}\bar{M}_2(j) = 0, \tag{6.3.7b}$$

$$A_K(i)\bar{N}_1(i) + \bar{N}_1(i)A_K'(i) + A_p(i)\bar{N}_1(i)A_p'(i) \\ - \gamma^{-2}\bar{M}_1(i)D(i)D'(i)\bar{N}_1(i) - \gamma^{-2}\bar{N}_1(i)\bar{M}_1(i)D(i)D'(i) + \pi_{ii}\bar{N}_1(i) \\ - \left(\gamma^{-2}D(i)D'(i)\bar{M}_2(i)\bar{N}_2(i) + \gamma^{-2}\bar{N}_2(i)\bar{M}_2(i)D(i)D'(i)\right) = 0 \tag{6.3.7c}$$

$$A_F(i)\bar{N}_2(i) + \bar{N}_2(i)A_F'(i) + A_p(i)\bar{N}_2(i)A_p'(i) + \pi_{ii}\bar{N}_2(i) + I_n = 0, \tag{6.3.7d}$$

$$R(i)K(i)\bar{N}_2(i) + B'(i)(\bar{M}_1(i)\bar{N}_1(i) + \bar{M}_2(i)\bar{N}_2(i)) = 0, \tag{6.3.7e}$$

where

$$F_\gamma(i) = -\gamma^{-2}D'(i)\bar{M}_1(i), A_K(i) = A(i) + B(i)K(i), A_F(i) \\ = A_K(i) - \gamma^{-2}D(i)D'(i)\bar{M}_1(i).$$

Suppose the system (6.3.3) is internal stable, then, the strategy set (v^*, u^*) with the form $v(t) = v^*(t) = \sum_{i=1}^{l} F_\gamma(i)\chi_{r_t=i}(t)x(t)$, $u(t) = u^*(t) = \sum_{i=1}^{l} K(i)\chi_{r_t=i}(t)x(t)$ is the stochastic H_2/H_∞ control set based on Stackelberg strategy.

Proof Given arbitrary $u(t) = K(r_t)x(t)$ of the leader, the follower facing the following optimization problem

$$\min_v J_1(x_0, i; v, K(r_t)x) \\ = \mathbf{E}\left\{\int_0^\infty \left[\gamma^2 v'(t)v(t) - x'(t)C'(r_t)C(r_t)x(t)\right]dt | r_0 = i\right\},$$

$$s.t.$$

$$dx(t) = [A_K(r_t)x(t) + D(r_t)v(t)]dt + A_p(r_t)x(t)dw(t).$$

From the conclusions of stochastic LQ problems, we can get the optimal feedback controller $v^\circ(t)$ is given by

$$v^\circ(t) = \sum_{i=1}^{l} F_\gamma(i)\chi_{r_t=i}(t)x(t) = -\gamma^{-2}\sum_{i=1}^{l} D'(i)\bar{M}_1(i)\chi_{r_t=i}(t)x(t), \tag{6.3.8}$$

where $\bar{M}_1(i)$ is the solution to

$$\mathsf{G}_1(\bar{M}_1(i), K(i)) = A_K'(i)\bar{M}_1(i) + \bar{M}_1(i)A_K(i) + A_p'(i)\bar{M}_1(i)A_p(i) - \gamma^{-2}\bar{M}_1(i)D'(i)D(i)\bar{M}_1(i) - C'(i)C(i) + \sum_{j=1}^{l}\pi_{ij}\bar{M}_1(j) = 0. \tag{6.3.9}$$

Therefore, equation (6.3.7a) holds.

On the other hand, if $A_F(i) = A_K(i) - \gamma^{-2}D(i)D'(i)\bar{M}_1(i)$ is asymptotically mean square stable, then the cost J_2 of the leader can be represented as

$$J_2\left(x_0, i; -\gamma^{-2}D'(r_t)\bar{M}_1(r_t)x, K(i)x\right) = \mathbf{Tr}(\bar{M}_2(i)), \tag{6.3.10}$$

where $\bar{M}_2(i)$ is the solution to

$$\mathsf{G}_2(\bar{M}_1(i), \bar{M}_2(i), K(i)) = A_F'(i)\bar{M}_2(i) + \bar{M}_2(i)A_F(i) + A_p'(i)\bar{M}_2(i)A_p(i) + Q(i) + K'(i)R(i)K(i) + \sum_{j=1}^{l}\pi_{ij}\bar{M}_2(j) = 0. \tag{6.3.11}$$

From (6.3.11) we know (6.3.7b) holds. Let us consider the following Lagrangian $\mathbf{H}$

$$\mathbf{H}(\bar{M}_1(i), \bar{M}_2(i), K(i)) = \mathbf{Tr}(\bar{M}_2(i)) + \mathsf{Tr}(\bar{N}_1(i)\mathsf{G}_1(\bar{M}_1(i), K(i))) + \mathsf{Tr}(\bar{N}_2(i)\mathsf{G}_2(\bar{M}_1(i), \bar{M}_2(i), K(i))), \tag{6.3.12}$$

where $\bar{N}_1(i)$ and $\bar{N}_2(i)$ are symmetric matrix of Lagrange multipliers.

As a necessary condition to minimization $\mathbf{Tr}(\bar{M}_2(i))$, we get

$$\frac{\partial \mathbf{H}}{\partial \bar{M}_1(i)} = A_K(i)\bar{N}_1(i) + \bar{N}_1(i)A_K'(i) + A_p(i)\bar{N}_1(i)A_p'(i) - \gamma^{-2}\bar{M}_1(i)D(i)D'(i)\bar{N}_1(i) - \gamma^{-2}\bar{N}_1(i)\bar{M}_1(i)D(i)D'(i) + \pi_{ii}\bar{N}_1(i) - \left(\gamma^{-2}D(i)D'(i)\bar{M}_2(i)\bar{N}_2(i) + \gamma^{-2}\bar{N}_2(i)\bar{M}_2(i)D(i)D'(i)\right) = 0, \tag{6.3.13a}$$

$$\frac{\partial \mathbf{H}}{\partial \bar{M}_1(i)} = A_F(i)\bar{N}_2(i) + \bar{N}_2(i)A_F'(i) + A_p(i)\bar{N}_2(i)A_p'(i) + \pi_{ii}\bar{N}_2(i) + I_n = 0, \tag{6.3.13b}$$

$$\frac{1}{2}\frac{\partial \mathbf{H}}{\partial K(i)} = R(i)K(i)\bar{N}_2(i) + B'(i)(\bar{M}_1(i)\bar{N}_1(i) + \bar{M}_2(i)\bar{N}_2(i)) = 0, \tag{6.3.13c}$$

Therefore, (6.3.7c)–(6.3.7e) hold.
This completes the proof of Theorem 6.3.1. □

6.4 Stochastic H_2/H_∞ Control to Markov Jump Singular System Based on Nash Games

6.4.1 Finite-Time Horizon Case

Consider the following continuous-time stochastic Markov jump singular systems with state-, control- and disturbance-dependent noise

$$\begin{cases} Edx(t) = [A_1(t,r_t)x(t) + B_1(t,r_t)u(t) + C_1(t,r_t)v(t)]dt \\ \qquad\qquad + [A_2(t,r_t)x(t) + B_2(t,r_t)u(t) + C_2(t,r_t)v(t)]dw(t), \\ z(t) = \begin{bmatrix} D(t,r_t)x(t) \\ F(t,r_t)u(t) \end{bmatrix}, \\ F'(t,r_t)F(t,r_t) = I, x(0) = x_0 \in \mathbb{R}^n, t \in [0,T]. \end{cases} \tag{6.4.1}$$

where $E \in \mathbb{R}^{n\times n}$ is a known singular matrix with $0 < rank(E) = k \leq n$, $x(t) \in \mathbb{R}^n$, $u(t) \in \mathbb{R}^{n_u}$, $v(t) \in \mathbb{R}^{n_v}$, and $z(t) \in \mathbb{R}^{n_z}$ are the system state, control input, exogenous input, and regulated output, respectively. All coefficients of (6.4.1) are assumed to be continuous matrix-valued functions of suitable dimensions.

For given disturbance attenuation $\gamma > 0$, define two associated performances as follows, $i \in \Xi$:

$$\begin{aligned} J_1(u,v;x_0,r_0) &= \gamma^2\|v(t)\|^2_{[0,T]} - \|z(t)\|^2_{[0,T]} \\ &= \mathbf{E}\left\{\int_0^T \left(\gamma^2\|v(t)\|^2 - \|z(t)\|^2\right)dt | r_0 = i\right\}, \end{aligned} \tag{6.4.2a}$$

$$J_2(u,v;x_0,r_0) = \|z(t)\|^2_{[0,T]} = \mathbf{E}\left\{\int_0^T \|z(t)\|^2 dt | r_0 = i\right\}. \tag{6.4.2b}$$

The definition of finite-time horizon stochastic H_2/H_∞ control problem is:

Definition 6.4.1 For system (6.4.1) and given $\gamma > 0$, $0 < T < \infty$, find, if possible, a state feedback control $u^*(t) \in L^2_{\mathcal{F}}(0,T;\mathbb{R}^{n_u})$ such that:

(i)
$$\|\mathcal{L}_T\| = \sup_{\substack{v \neq 0 \\ u = u^* \\ x_0 = 0}} \frac{\mathbf{E}\left\{\int_0^T \left(\|D(t,r_t)x(t)\|^2 + \|u^*(t)\|^2\right)dt | r_0 = i\right\}^{1/2}}{\mathbf{E}\left\{\int_0^T \|v(t)\|^2 dt | r_0 = i\right\}^{1/2}} < \gamma, \tag{6.4.3}$$

where $i \in \Xi$, and $\mathcal{L}_T$ is an operator associated with system (6.4.1) which is defined as

$$\mathcal{L}_T : L^2_{\mathcal{F}}(0,T;\mathbb{R}^{n_v}) \mapsto L^2_{\mathcal{F}}(0,T;\mathbb{R}^{n_z}),$$
$$\mathcal{L}_T(v(t)) = z(t)|_{x_0=0}, t \in [0,T].$$

(ii) When the worst case disturbance $v^*(t) \in L^2_{\mathcal{F}}(0,T;\mathbb{R}^{n_v})$, if it exists, is applied to (6.4.1), $u^*(t)$ minimizes the output energy

$$J_2(u,v^*;x_0,r_0) = \mathbf{E}\left\{\int_0^T \left(\|D(t,r_t)x(t)\|^2 + \|u(t)\|^2\right)dt|r_0 = i\right\}, \quad (6.4.4)$$

where $v^*(t)$ is defined as

$$v^*(t) = \arg\min\left\{J_1(u^*,v;x_0,r_0) = \mathbf{E}\left\{\int_0^T \left(\gamma^2\|v(t)\|^2 - \|z(t)\|^2\right)dt|r_0 = i\right\}\right\}.$$

If the above (u^*,v^*) exist, then we say that the finite-time horizon H_2/H_∞ control of system (6.4.1) is solvable and has a pair of solutions (u^*,v^*).

In other words, for given two cost functions defined in (6.4.2a) and (6.4.2b), the finite-time horizon H_2/H_∞ control of system (6.4.1) is equivalent to finding the Nash equilibrium (u^*,v^*), such that

$$\begin{gathered} J_1(u^*,v^*;x_0,r_0) \le J_1(u^*,v;x_0,r_0), J_2(u^*,v^*;x_0,r_0) \le J_2(u,v^*;x_0,r_0), \\ (u(t),v(t)) \in L^2_{\mathcal{F}}(0,T;\mathbb{R}^{n_u}) \times L^2_{\mathcal{F}}(0,T;\mathbb{R}^{n_v}), r_0 \in \Xi. \end{gathered} \quad (6.4.5)$$

Clearly, if the Nash equilibrium (u^*,v^*) exist, u^* is our desired H_2/H_∞ controller, and v^* is the worst case disturbance. In this case, we also say that the stochastic H_2/H_∞ control admits a pair of solutions (u^*,v^*).

Before giving the main results, some preliminary work needs to be introduced. Consider the following stochastic singular perturbed system with Markov jump parameters

$$\begin{cases} Edx(t) = [A_1(t,r_t)x(t) + C_1(t,r_t)v(t)]dt + [A_2(t,r_t)x(t) + C_2(t,r_t)v(t)]dw(t), \\ z(t) = D(t,r_t)x(t), \quad x(0) = x_0 \in \mathbb{R}^n, \quad t \in [0,T]. \end{cases} \quad (6.4.6)$$

For any given $0 < T < \infty$, associated with system (6.4.6), the perturbation operator $\tilde{\mathcal{L}}_T : L^2_{\mathcal{F}}(0,T;\mathbb{R}^{n_v}) \mapsto L^2_{\mathcal{F}}(0,T;\mathbb{R}^{n_z})$ is defined as $\tilde{\mathcal{L}}_T(v(t)) = z(t)|_{x_0=0} = D(t,r_t)x(t)|_{x_0=0}, t \in [0,T]$, then

$$\|\tilde{\mathcal{L}}_T\| = \sup_{v\neq 0, x_0=0} \frac{\mathbf{E}\left\{\int_0^T \|D(t,r_t)x(t)\|^2 dt | r_0 = i\right\}^{1/2}}{\mathbf{E}\left\{\int_0^T \|v(t)\|^2 dt | r_0 = i\right\}^{1/2}}.$$

In our subsequent analysis, we define $M_i = M(t,i)$, $M_{1i} = M_1(t,i)$, $M_{2i} = M_2(t,i)$, $i \in \Xi$ for convenience.

Lemma 6.4.1 *For system (6.4.6) and given disturbance attenuation $\gamma > 0$, $\|\tilde{\mathcal{L}}_T\| < \gamma$ iff the following constrained equations*

$$\begin{cases} E'\dot{P}_iE + E'P_iA_{1i} + A'_{1i}P_iE + A'_{2i}P_iA_{2i} - D'_iD_i + \sum\limits_{j=1}^{l} \pi_{ij}E'P_jE \\ -(EP_iC_{1i} + A'_{2i}P_iC_{2i})\left(\gamma^2 I + C'_{2i}P_iC_{2i}\right)^{-1}\left(C'_{1i}P_iE + C'_{2i}P_iA_{2i}\right) = 0, \\ E'P(T,i)E = 0, \\ \gamma^2 I + C'_{2i}P_iC_{2i} > 0, \forall t \in [0,T], \quad i \in \Xi, \end{cases} \tag{6.4.7}$$

have a bounded solution $P(t) = (P(t,1),\ldots,P(t,l)) \leq 0 \in \mathcal{C}([0,T];\mathcal{S}_l^n)$.

Proof The details are similar with Ref. [1], so we omitted it here.

The following theorem presents the main results of finite-time horizon stochastic H_2/H_∞ control.

Theorem 6.4.1 *For system (6.4.1), given a disturbance attenuation level $\gamma > 0$ and $0 < T < \infty$, the stochastic H_2/H_∞ control admits a pair of solutions (u^*, v^*) with*

$$u^*(t) = \sum_{i=1}^{l} K_2(t,i)\chi_{r_t=i}(t)x(t),\ v^*(t) = \sum_{i=1}^{l} K_1(t,i)\chi_{r_t=i}(t)x(t), \tag{6.4.8}$$

if and only if for $\forall t \in [0,T]$, $i \in \Xi$, the following four coupled differential Riccati equations

$$\begin{cases} E'\dot{P}_i^1E + E'P_i^1\tilde{A}_{1i} + \tilde{A}'_{1i}P_i^1E + \tilde{A}'_{2i}P_i^1\tilde{A}_{2i} + \tilde{Q}(i) + \sum\limits_{j=1}^{l} \pi_{ij}E'P_j^1E \\ -(E'P_i^1C_{1i} + \tilde{A}'_{2i}P_i^1C_{2i})\left(\gamma^2 I + C'_{2i}P_i^1C_{2i}\right)^{-1}\left(C'_{1i}P_i^1E + C'_{2i}P_i^1\tilde{A}_{2i}\right) = 0, \\ E'P^1(T,i)E = 0, \\ \gamma^2 I + C'_{2i}P_i^1C_{2i} > 0, \ i \in \Xi. \end{cases} \tag{6.4.9a}$$

$$K_{1i} = -\left(\gamma^2 I + C'_{2i}P_i^1C_{2i}\right)^{-1}\left(C'_{1i}P_i^1E + C'_{2i}P_i^1\tilde{A}_{2i}\right). \tag{6.4.9b}$$

$$\begin{cases} E'\dot{P}_i^2E+E'P_i^2\bar{A}_{1i}+\bar{A}_{1i}'P_i^2E+\bar{A}_{2i}'P_i^2\bar{A}_{2i}+D_i'D_i+\sum_{j=1}^{l}\pi_{ij}E'P_j^2E \\ -\left(E'P_i^2B_{1i}+\bar{A}_{2i}'P_i^2B_{2i}\right)\left(I+B_{2i}'P_i^2B_{2i}\right)^{-1}\left(B_{1i}'P_i^2E+B_{2i}'P_i^2\bar{A}_{2i}\right)=0, \\ E'+B_{2i}'P_i^2B_{2i}>0,\ i\in\Xi. \\ I+B_{2i}'P_i^2B_{2i}>0,\ i\in\Xi. \end{cases} \tag{6.4.9c}$$

$$K_{2i}=-\left(I+B_{2i}'P_i^2B_{2i}\right)^{-1}\left(B_{1i}'P_i^2E+B_{2i}'P_i^2\bar{A}_{2i}\right), \tag{6.4.9d}$$

where

$$\begin{aligned} &\tilde{A}_{1i}=A_{1i}+B_{1i}K_{2i},\tilde{A}_{2i}=A_{2i}+B_{2i}K_{2i},\tilde{Q}_i=-(D_i'D_i+K_{2i}'K_{2i}), \\ &\bar{A}_{1i}=A_{1i}+C_{1i}K_{1i},\bar{A}_{2i}=A_{2i}+C_{2i}K_{1i} \end{aligned}$$

admit solutions $(P^1(t),P^2(t),K_1(t),K_2(t))$,

$$\begin{aligned} P^1(t)&=(P^1(t,1),\cdots,P^1(t,l))\le 0\in\mathcal{C}([0,T];\mathcal{S}_l^n), \\ P^2(t)&=(P^2(t,1),\cdots,P^2(t,l))\ge 0\in\mathcal{C}([0,T];\mathcal{S}_l^n). \end{aligned}$$

Proof Please refer to the proof of Theorem 6.2.1, we don't give it in detail here. □

6.4.2 Infinite-Time Horizon Case

6.4.2.1 Preliminaries

Consider the following controlled stochastic singular systems with Markovian jumps

$$Edx(t)=[A(r_t)x(t)+B(r_t)u(t)]dt+A_1(r_t)x(t)dw(t) \tag{6.4.10}$$

where $E\in\mathbb{R}^{n\times n}$ is a known singular matrix with $0<rank(E)=k\le n$, $x(t)\in\mathbb{R}^n$ is the system state, $u(t)\in\mathbb{R}^{n_u}$ is control input, all coefficient matrices are assumed to be constant with compatible dimensions for given $r_t=i\in\Xi$.

For system (6.4.10), applying generalized Itô's formula to $x'E'P(i)Ex$, we have

Lemma 6.4.2 *Suppose* $P=(P(1),P(2),\cdots,P(l))\in\mathcal{S}_l^n$ *is given, then for system (6.4.10) with initial state* $(x_0,i)\in\mathbb{R}^n\times\Xi$, *we have for any* $T>0$,

$$\mathbf{E}\left\{\int_0^T [x'(t)(E'P(r_t)A(r_t)+A'(r_t)P(r_t)E+A_1'(r_t)P(r_t)A_1(r_t)+\sum_{j=1}^{l}\pi_{r_tj}E'P(j)E)x(t)\right.$$
$$\left.+2u'(t)B'(r_t)P(r_t)Ex(t)]dt|r_0=i\right\}=\mathbf{E}[x'(T)E'P(r_T)Ex(T)|r_0=i]-x_0'E'P(i)Ex_0. \tag{6.4.11}$$

6.4.2.2 Main Results

For simplicity and without loss of generality, consider the following Markov jump singular systems described by stochastic differential equation

$$\begin{cases} Edx(t)=[A(r_t)x(t)+B_2(r_t)u(t)+B_1(r_t)v(t)]dt+[A_1(r_t)x(t)+C_2(r_t)u(t)]dw(t), \\ z(t)=\begin{bmatrix} C_0(r_t)x(t) \\ D(r_t)u(t)\end{bmatrix}, D'(i)D(i)\equiv I, x(0)=x_0\in\mathbb{R}^n, i\in\Xi. \end{cases} \tag{6.4.12}$$

where $x(t)\in\mathbb{R}^n$, $u(t)\in\mathbb{R}^{n_u}$, $v(t)\in\mathbb{R}^{n_v}$, and $z(t)\in\mathbb{R}^{n_z}$ are the system state, control input, exogenous input, and regulated output, respectively. All coefficients are assumed to be constants. Define two associated cost functions as follows:

$$J_1^\infty(u,v;x_0,i)=\mathbf{E}\left\{\int_0^\infty\left[\gamma^2\|v(t)\|^2-\|z(t)\|^2\right]dt|r_0=i\right\}. \tag{6.4.13a}$$

$$J_2^\infty(u,v;x_0,i)=\mathbf{E}\left\{\int_0^\infty\|z(t)\|^2dt|r_0=i\right\},\quad i\in\Xi. \tag{6.4.13b}$$

The infinite-time horizon stochastic H_2/H_∞ control problem of system (6.4.12) is described as follows.

Definition 6.2.4 [6] For given disturbance attenuation level $\gamma>0$, if we can find $u^*(t)\times v^*(t)\in L^2_{\mathcal{F}}(0,\infty;\mathbb{R}^{n_u})\times L^2_{\mathcal{F}}(0,\infty;\mathbb{R}^{n_v})$, such that

(i) $u^*(t)$ stabilizes system (6.4.12) internally, i.e. when $v(t)=0$, $u=u^*$, the state trajectory of (6.4.11) with any initial value $(x_0,i)\in\mathbb{R}^n\times\Xi$ satisfies

$$\lim_{t\to\infty}\mathbf{E}\left[\|x(t)\|^2|r_0=i\right]=0.$$

(ii) $|\mathcal{L}_{u^*}|_\infty<\gamma$ with

$$\|\mathcal{L}_{u^*}\|_\infty = \sup_{\substack{v \in L^2_{\mathcal{F}}(0,\infty;\mathbb{R}^{n_v}) \\ v \neq 0, u = u^*, x_0 = 0}} \frac{\mathbf{E}\left\{\int_0^\infty \left[\|C_0(r_t)x(t)\|^2 + \|u^*(t)\|^2\right] | r_0 = i\right\}^{1/2}}{\mathbf{E}\left\{\int_0^\infty \|v(t)\|^2 dt | r_0 = i\right\}^{1/2}}.$$

(iii) When the worst case disturbance $v^*(t) \in L^2_{\mathcal{F}}(0,\infty;\mathbb{R}^{n_v})$, if existing, is applied to (6.4.12), $u^*(t)$ minimizes the output energy

$$J_2^\infty(u, v^*; x_0, i) = \mathbf{E}\left\{\int_0^\infty \left[\|C_0(r_t)x(t)\|^2 + \|u(t)\|^2\right] dt | r_0 = i\right\}, \quad i \in \Xi,$$

where $v^*(t)$ is defined as

$$v^*(t) = \arg\min\left\{J_1^\infty(u^*, v; x_0, i) = \mathbf{E}\left\{\int_0^\infty \left(\gamma^2\|v(t)\|^2 - \|z(t)\|^2\right) dt | r_0 = i\right\}\right\}.$$

If the above (u^*, v^*) exist, then we say that the infinite-time horizon H_2/H_∞ control of system (6.4.12) is solvable and has a pair of solutions. Obviously, (u^*, v^*) is the Nash equilibrium strategies such that

$$J_1^\infty(u^*, v^*; x_0, i) \le J_1^\infty(u^*, v; x_0, i), \tag{6.4.14}$$

$$J_2^\infty(u^*, v^*; x_0, i) \le J_2^\infty(u, v^*; x_0, i), \quad i \in \Xi. \tag{6.4.15}$$

The main result of the infinite-time horizon stochastic H_2/H_∞ control is presented by the following theorem, which can be shown following the line of Theorem 2.1 and Theorem 1 presented in Ref. [4].

Theorem 6.4.2 *For a given disturbance attenuation $\gamma > 0$, suppose systems (6.4.12) is stochastically stablizable, infinite-time horizon stochastic H_2/H_∞ control has a pair of solutions (u^*, v^*) with $u^*(t) = K_2(r_t)x(t)$ and $v^*(t) = K_1(r_t)x(t)$, where $K_2(i) \in \mathcal{M}^l_{n_u,n}$ and $K_1(i) \in \mathcal{M}^l_{n_v,n}$ are constant matrices, if and only if the following four coupled algebraic Riccati equations (6.4.16a)–(6.4.16d) admit solutions (P_1, P_2, K_1, K_2) as $P_1 = (P_1(1), P_1(2), \cdots, P_1(l)) \le 0 \in \mathcal{S}^n_l, P_2 = (P_2(1), P_2(2), \cdots, P_2(l)) \ge 0 \in \mathcal{S}^n_l$*

$$\begin{aligned} &E'P_1(i)\tilde{A}(i) + \tilde{A}'(i)P_1(i)E + \tilde{A}_1'(i)P_1(i)\tilde{A}_1(i) + \tilde{Q}(i) \\ &+ \sum_{j=1}^{l} \pi_{ij}E'P_1(j)E - \gamma^{-2}E'P_1(i)B_1(i)B_1'(i)P_1(i)E = 0, \; i \in \Xi. \end{aligned} \tag{6.4.16a}$$

$$K_1(i) = -\gamma^{-2}B_1'(i)P_1(i)E. \tag{6.4.16b}$$

$$\begin{cases} E'P_2(j)\bar{A}(j) + \bar{A}'(j)P_2(j)E + A_1'(j)P_2(j)A_1(j) + C_0'(j)C_0(j) + \sum_{k=1}^{l} \pi_{jk}E'P_2(k)E \\ -\left(E'P_2(j)B_2(j) + A_1'(j)P_2(j)C_2(j)\right)\left(I + C_2'(j)P_2(j)C_2(j)\right)^{-1} \\ \times\left(B_2'(j)P_2(j)E + C_2'(j)P_2(j)A_1(j)\right) = 0, \\ I + C_2'(j)P_2(j)C_2(j) > 0,\ j \in \Xi. \end{cases} \tag{6.4.16c}$$

$$K_2(j) = -\left(I + C_2'(j)P_2(j)C_2(j)\right)^{-1}\left(B_2'(j)P_2(j)E + C_2'(j)P_2(j)A_1(j)\right), \tag{6.4.16d}$$

where

$$\tilde{A} = A - B_2K_2, \tilde{A}_1 = A_1 - C_2K_2, \tilde{Q} = -(C_0'C_0 + K_2'K_2), \bar{A} = A - B_1K_1.$$

Proof Please refer to the proof of Theorem 6.2.1, we don't give it in detail here. □

6.5 Stochastic H$_2$/H$_\infty$ Control to Markov Jump Singular Systems Based on Stackelberg Game

6.5.1 Preliminary Results

Consider the following Markov jump singular systems

$$\begin{cases} Edx(t) = [A(r_t)x(t) + D(r_t)v(t)]dt + A_p(r_t)x(t)dw(t), \\ z(t) = C(r_t)x(t), \quad x(0) = x_0 \in \mathbb{R}^n. \end{cases} \tag{6.5.1}$$

where $v(t) \in \mathbb{R}^{n_v}$ and $z(t) \in L^2_{\mathcal{F}}(0, \infty; \mathbb{R}^{n_z})$ are, respectively, disturbance signal and controlled output. For system (6.5.1), the perturbed operator $\mathcal{L}$: $L^2_{\mathcal{F}}(0, \infty; \mathbb{R}^{n_v}) \mapsto L^2_{\mathcal{F}}(0, \infty; \mathbb{R}^{n_z})$ of $\mathcal{L}(v(t)) = z(t)|_{x_0=0} = C(r_t)x(t)|_{x_0=0}$ is defined by

$$\|\mathcal{L}\|_\infty = \sup_{v \neq 0, x_0=0} \frac{\mathbf{E}\left\{\int_0^\infty \|z(t)\|^2 dt | r_0 = i\right\}^{1/2}}{\mathbf{E}\left\{\int_0^\infty \|v(t)\|^2 dt | r_0 = i\right\}^{1/2}}.$$

The following lemma extends the real bounded lemma of Markov jump linear systems proposed by Huang and Zhang [3] to Markov jump singular systems:

Lemma 6.5.2 *Given a disturbance attenuation level $\gamma > 0$, the stochastic system (6.5.1) is internal stable and $\|\mathcal{L}\|_\infty < \gamma$ for some γ, if and only if the following equation*

$$\begin{aligned} & A'(i)P(i)E + E'P(i)A(i) + A_p'(i)P(i)A_p(i) + C'(i)C(i) \\ & + \gamma^{-2}E'P(i)D(i)D'(i)P(i)E + \sum_{j=1}^{l} \pi_{ij}E'P(j)E = 0 \end{aligned} \tag{6.5.2}$$

has a solution $P = (P(1), \ldots, P(l)) > 0 \in \mathcal{S}_l^n$.

In this case, $v^*(t) = \sum_{i=1}^{l} F^*(i)\chi_{r_t=i}(t)x(t) = \gamma^{-2}\sum_{i=1}^{l} D'(i)P(i)\chi_{r_t=i}(t)Ex(t)$ is the worst disturbance of the system.

6.5.2 Problem Formulation

Consider the following controlled stochastic Markov jump singular systems

$$\begin{cases} Edx(t) = [A(r_t)x(t) + B(r_t)u(t) + D(r_t)v(t)]dt + A_p(r_t)x(t)dw(t), \\ z(t) = C(r_t)x(t), x(0) = x_0. \end{cases} \tag{6.5.3}$$

where $x(t) \in \mathbb{R}^n$, $u(t) \in \mathbb{R}^{n_u}$, $v(t) \in \mathbb{R}^{n_v}$, and $z(t) \in \mathbb{R}^{n_z}$ are, respectively, the system state, control input, disturbance signal and controlled output. $E \in \mathbb{R}^{n\times n}$ is a known singular matrix with $0 < rank(E) = k \leq n$. In (6.5.3), $A(r_t) = A(i)$, $B(r_t) = B(i)$, $C(r_t) = C(i)$, $D(r_t) = D(i)$ and $A_p(r_t) = A_p(i)$, when $r_t = i$, while $A(i)$, etc., $i = 1, \ldots, l$, are constant matrices with suitable size.

The definition of stochastic H_2/H_∞ control based on Stackelberg game is given below:

Definition 6.5.1 [6] If there exists a strategy set (v^*, u^*), such that

$$J_2(x_0, i; v^*, u^*) \leq J_2(x_0, i; v^o(u), u), \quad \forall u \in \mathbb{R}^{n_u}, \tag{6.5.4}$$

where

$$J_1(x_0, i; v^o(u), u) = \min_v J_1(x_0, i; v, u), \tag{6.5.5}$$

and

$$v^* = v^o(u^*), \tag{6.5.6a}$$

$$J_1(x_0, i; v, u) = \mathbf{E}\left\{\int_0^\infty \left[\gamma^2\|v(t)\|^2 - \|z(t)\|^2\right]dt | r_0 = i\right\}, \tag{6.5.6b}$$

$$J_2(x_0, i; v, u) = \mathbf{E}\left\{ \int_0^{\infty} [x'(t)Q(r_t)x(t) + u'(t)R(r_t)u(t)]dt | r_0 = i \right\}, \quad (6.5.6c)$$
$$Q(r_t) = Q'(r_t) \geq 0, \quad R(r_t) = R'(r_t) > 0.$$

then, this strategy set is the desired stochastic H_2/H_∞ control set.

Without loss of generality, we restrict our strategies to linear state feedback case, i.e., the closed-loop stochastic H_2/H_∞ sets have the following form

$$v(t) = \sum_{i=1}^{l} F_\gamma(i)\chi_{r_t=i}(t)x(t), u(t) = \sum_{i=1}^{l} K(i)\chi_{r_t=i}(t)x(t). \quad (6.5.7)$$

6.5.3 Main Results

The following theorem presents the main results of stochastic H_2/H_∞ control based on Stackelberg strategy:

Theorem 6.5.1 *Suppose that the following cross-coupled matrix-valued equations (6.5.8a)–(6.5.8e) has solutions* $\bar{M}_1(i) \leq 0$, $\bar{M}_2(i) \geq 0$, $\bar{N}_j(i), j = 0, 1$, $F_\gamma(i)$ *and* $K(i)$, $i = 1, \ldots l$

$$\begin{aligned} &A_K'(i)E\bar{M}_1(i) + \bar{M}_1(i)E'A_K(i) + A_p'(i)\bar{M}_1(i)A_p(i) \\ &- \gamma^{-2}E'\bar{M}_1(i)D'(i)D(i)\bar{M}_1(i)E - C'(i)C(i) + \sum_{j=1}^{l} \pi_{ij}E'\bar{M}_1(j)E = 0, \end{aligned} \quad (6.5.8a)$$

$$\begin{aligned} &A_F'(i)E\bar{M}_2(i) + \bar{M}_2(i)E'A_F(i) + A_p'(i)\bar{M}_2(i)A_p(i) + Q(i) \\ &+ K'(i)R(i)K(i) + \sum_{j=1}^{l} \pi_{ij}E'\bar{M}_2(j)E = 0, \end{aligned} \quad (6.5.8b)$$

$$\begin{aligned} &A_K(i)E\bar{N}_1(i) + \bar{N}_1(i)E'A_K'(i) + A_p(i)\bar{N}_1(i)A_p'(i) \\ &- \gamma^{-2}E'\bar{M}_1(i)D(i)D'(i)E\bar{N}_1(i) - \gamma^{-2}\bar{N}_1(i)E'\bar{M}_1(i)D(i)D'(i)E + \pi_{ii}E'\bar{N}_1(i)E \\ &- \left(\gamma^{-2}E'D(i)D'(i)\bar{M}_2(i)E\bar{N}_2(i) + \gamma^{-2}\bar{N}_2(i)E'\bar{M}_2(i)D(i)D'(i)E\right) = 0, \end{aligned} \quad (6.5.8c)$$

$$A_F(i)E\bar{N}_2(i) + \bar{N}_2(i)E'A_F'(i) + A_p(i)\bar{N}_2(i)A_p'(i) + \pi_{ii}E'\bar{N}_2(i)E + I_n = 0, \quad (6.5.8d)$$

$$R(i)K(i)\bar{N}_2(i) + B'(i)(\bar{M}_1(i)E'\bar{N}_1(i) + \bar{M}_2(i)E'\bar{N}_2(i)) = 0, \quad (6.5.8e)$$

where

$$F_\gamma(i) = -\gamma^{-2}D'(i)\bar{M}_1(i)E, A_K(i) = A(i) + B(i)K(i),$$
$$A_F(i) = A_K(i) - \gamma^{-2}D(i)D'(i)\bar{M}_1(i)E.$$

Suppose the system (6.5.3) is internal stable, then, the strategy set (v^*, u^*) *with the form* $v(t) = v^*(t) = \sum_{i=1}^{l} F_\gamma(i)\chi_{r_t=i}(t)x(t), \quad u(t) = u^*(t) = \sum_{i=1}^{l} K(i)\chi_{r_t=i}(t)x(t)$, *is the stochastic* H_2/H_∞ *control set based on Stackelberg strategy.*

Proof The proof of this theorem can be referring to the proof step of Theorem 6.3.1 to draw, and we don't give it in detail here. □

6.6 Summary

This chapter deals with the stochastic H_2/H_∞ control problems of continuous-time Markov jump linear systems and Markov jump singular systems. The main methodology used in this chapter is the game theory approach, by introducing two players, wherein the control designer of the system is considered as P_1, the exogenous disturbance is considered as "nature" P_2, the stochastic H_2/H_∞ control problems can be converted into a two person nonzero-sum stochastic differential games, and then, using the relevant results of stochastic differential games for Markov jump linear systems and Markov jump singular systems obtained in Chaps. 3 and 5, the necessary and sufficient condition for the existence of stochastic H_2/H_∞ control is equivalent to the corresponding matrixed-value differential (algebraic) equations have solutions, and meanwhile, the explicit mathematical expressions of the stochastic H_2/H_∞ control is given. The conclusion of this chapter not only enriched the existing results of robust control of stochastic systems, but also widened the differential game method in handling various control problems, and has laid a theoretical foundation for later study in this book.

References

1. Zhang, W., & Chen, B. S. (2006). State feedback H_∞ control for a class of nonlinear stochastic systems. *SIAM Journal on Control and Optimization, 44*(6), 1973–1991.
2. Boukas, E. K., & Liu, Z. K. (2001). Jump linear quadratic regulator with controlled jump rates. *IEEE Transactions on Automatic Control, 46*(2), 301–305.
3. Huang, Y., Zhang, W., & Feng, G. (2008). Infinite horizon H_2/H_∞ control for stochastic systems with Markovian jumps. *Automatica, 44*(3), 857–863.
4. Chen, B. S., & Zhang, H. (2004). Stochastic H_2/H_∞ control with state-dependent noise. *IEEE Transactions on Automatic Control, 49*(1), 45–57.

5. Li, X., Zhou, X. Y., & Rami, M. A. (2003). Indefinite Stochastic linear quadratic control with Markovian jumps in infinite time horizon. *Journal of Global Optimization, 27*(2–3), 149–175.
6. Medanic, J. (1978). Closed-loop Stackelberg strategies in linear-quadratic problems. *IEEE Transactions on Automatic Control, 23*(4), 632–637.

Chapter 7
Applications of Stochastic Differential Game Theory for Markov Jump Linear Systems to Finance and Insurance

This chapter mainly introduces applications of stochastic differential game theory for Markov jump linear systems to finance and insurance. Firstly, a risk minimization problem is considered in a continuous-time Markovian regime switching financial model modulated by a continuous-time, finite-state, Markov chain. And then, European option valuation under Markovian regime-switching models is studied. Lastly, a game theoretic approach for optimal investment-reinsurance problem of an insurance company under Markovian regime-switching models is introduced in this chapter.

7.1 Introduction

In recent years, Markovian regime-switching models have attracted much attention by researchers and practitioners in economics and finance. Econometric applications of Markovian regime-switching were pioneered by the original work of Reinhard (1984) in which different states of the Markovian chain represent different stages of the economic state, known as the risk model with Markov-modulation by Asmussen (1989) [1]. The Markov-modulation can explain changes in macroeconomic conditions, changes in political systems, influence of major financial news, different stages of business cycles and so on. Presently, portfolio selection and option pricing models with Markov-modulation have been discussed by many researcher, and this has been an important problem from both theoretical and practical perspectives.

Moreover, game theory reflects rational thinking modes of players, which, especially stochastic differential game, has been an important method for economic analyzation [2, 3]. So, by means of stochastic differential game, this chapter discusses problems of portfolio risk minimization, option pricing and optimal investment of an insurance company under Markovian regime-switching models. Considering the market as a "virtual" game player, a two-player, zero-sum, stochastic differential game

C.-k. Zhang et al., *Non-cooperative Stochastic Differential Game Theory of Generalized Markov Jump Linear Systems*, Studies in Systems, Decision and Control 67, DOI 10.1007/978-3-319-40587-2_7

model between investors and markets is built. A verification theorem for the Hamilton_Jacobi_Bellman (HJB) solution of the game is provided.

7.2 Portfolio Risk Minimization and Differential Games

Risk management is an important issue in the modern banking and finance industries. Some recent financial crises, including the Asian financial crisis, the collapse of Long-Term Capital Management, the turmoil at Barings and Orange Country, raise the concern of regulators about the risk taking activities of banks and financial institutions and their practice of risk management. Recently, Value at Risk (VaR) has emerged as a standard and popular tool for risk measurement and management. VaR tells us the extreme loss of a portfolio over a fixed time period at a certain probability (confidence) level. Artzner et al. [4] develop a theoretical approach for developing measures of risk. They present a set of four desirable properties for measures of risk and introduce the class of coherent risk measures. They point out that VaR does not, in general, satisfy one of the four properties, namely, the subadditivity property. This motivates the quest for some theoretically consistent risk measures. Föllmer and Schied [5] argue that the risk of a portfolio might increase nonlinearly with the portfolio's size due to the liquidity risk. They relax the subadditive and positive homogeneous properties and replace them with the convex property. They introduce the class of convex risk measures, which include the class of coherent risk measures. Elliott and Kopp [6] provide a comprehensive account of coherent risk measures and convex risk measures.

In the past two decades or so, applications of regime-switching models in finance have received much attention. However, relatively little attention has been paid to the use of regime-switching models for quantitative risk management until recently. It is important to take the regime-switching effect into account in long-term financial risk management, such as managing the risk of pension funds, since there might be structural changes in the economic fundamentals over a long time period. Some recent works concerning the regime-switching effect on quantitative risk measurement include [7, 8], and others. However, these works mainly concern certain aspects of quantitative risk measurement and do not focus on risk management and control issues.

In this note, we explore the state of the art of a stochastic differential game for minimizing portfolio risk under a continuous-time Markovian regime-switching financial model. Stochastic differential games are an important topic in both mathematics and economics. Some early works on the mathematical theory of stochastic differential games include [9, 10], and others. Some recent works on stochastic differential games and their applications include [11–14], and others. Here, we suppose that an investor can only invest in a money market account and a stock whose price process follows a Markovian regime-switching geometric Brownian motion (GBM). The interest rate of the money market account, the drift and the volatility of the stock are modulated by a continuous-time, finite-state,

Markov chain. The states of the chain are interpreted as different states of an economy. For example, they may be interpreted as the credit ratings of a region, or a sovereign. They may also be interpreted as proxies of the levels of some observable (macro)-economic indicators, such as gross domestic product and retail price index. The Markovian regime-switching model provides a natural way to describe the impact of structural changes in (macro)-economic condition on asset price dynamics and the stochastic evolution of investment opportunity sets. We adopt a particular form of convex risk measure introduced by Föllmer and Schied, which includes the entropic risk measure as a special case. The entropic risk measure is a typical example of convex risk measure and corresponds to an exponential utility function, (see Barrieu and El-Karoui [15, 16]). Our goal is to minimize the convex risk measure of the terminal wealth of the investor. Following the plan of Mataramvura and Øksendal [17], we formulate the problem into a Markovian regime-switching stochastic differential game with two players, namely, the investor and the market.

In our model, the investor faces two sources of risk, namely, the diffusion risk due to fluctuations of financial prices and the regime-switching risk due to the change in the (macro)-economic condition. Here, we take into account these two sources of risk in evaluating and controlling the risk the investor faces. To achieve this, we introduce a product of two density processes, one for the Brownian motion and one for the Markov chain process, to generate a family of real-world probability measures in the representation of the convex risk measure. So, the market has two control variables, namely, the market price of risk for the change of measures related to the Brownian motion and the rate matrix of the Markov chain. We provide a verification theorem for the Markovian regime-switching HJB equation to the solution of the game corresponding to the risk minimization problem.

This note is based on part of [18]. We state results which will be published later in [18] without proofs.

7.2.1 Asset Price Dynamics

We consider a continuous-time financial model consisting of two primitive assets, namely, a money market account and a stock. These assets are assumed to be tradable continuously on a fixed time horizon $\mathcal{T} := [0, T]$, where $T \in (0, \infty)$. We fix a complete probability space $(\Omega, \mathcal{F}, \mathcal{P})$, where $\mathcal{P}$ represents a reference probability measure from which a family of absolutely continuous real-world probability measures will be generated.

Now, we introduce a continuous-time, finite-state, Markov chain to describe the evolution of the states of an economy over time. Throughout the note, we use boldface letters to denote vectors, or matrices. Let $\mathbf{X} := \{\mathbf{X}(t)\}_{t \in \mathcal{T}}$ denote a continuous-time, finite-state, Markov chain on $(\Omega, \mathcal{F}, \mathcal{P})$ with a finite state space $\mathcal{S} := \{\mathbf{s}_1, \mathbf{s}_2, \ldots, \mathbf{s}_N\}$. The states of the chain represent different state of the economy. Without loss of generality, we identify the state space of the chain to be a finite set of unit vectors $\mathcal{E} := \{\mathbf{e}_1, \mathbf{e}_2, \ldots, \mathbf{e}_N\}$, where $\mathbf{e}_i \in \mathbb{R}^N$ and the jth component of $\mathbf{e}_i$ is the Kronecker delta δ_{ij}, for each $i, j = 1, 2, \ldots, N$. $\mathbf{e}_i$ is called the canonical state space of $\mathbf{X}$.

Let $\mathbf{A}(t) = [a_{ij}(t)]_{i,j=1,2,\ldots,N}, t \in \mathcal{T}$, denote a family of generators, or rate matrices, of the chain $\mathbf{X}$ under $\mathcal{P}$. Here, $a_{ij}(t)$ represents the instantaneous intensity of the transition of the chain $\mathbf{X}$ from state i to state j at time t. Note that for each $t \in \mathcal{T}$ and $a_{ij}(t) \geq 0$ $(i \neq j)$, $\sum_{i=1}^{N} a_{ij}(t) = 0$, so $a_{ii}(t) \leq 0$. We assume that for each $i, j = 1, 2, \ldots, N$ and each $t \in \mathcal{T}$, $a_{ij}(t) > 0$. With the canonical representation of the state space of the chain, Elliott et al. [19] provide the following semimartingale decomposition for $\mathbf{X}$:

$$\mathbf{X}(t) = \mathbf{X}(0) + \int_0^t \mathbf{A}(u)\mathbf{X}(u)du + \mathbf{M}(t)$$

where $\{\mathbf{M}(t)\}_{t \in \mathcal{T}}$ is an $\mathbb{R}^N$-valued martingale with respect to the filtration generated by $\mathbf{X}$ under $\mathcal{P}$.

Let y' denote the transpose of a vector, or a matrix y. $\langle .,. \rangle$ is the scalar product in $\mathbb{R}^N$. The instantaneous market interest rate $r(t)$ of the money market account B is determined by the Markov chain as:

$$r(t) = \langle \mathbf{r}, \mathbf{X}(t) \rangle$$

where $\mathbf{r} := (r_1, r_2, \ldots, r_N)' \in \mathbb{R}^N$ with $r_i > 0$ for each $i = 1, 2, \ldots, N$. Then, the evolution of the balance of the money market account follows:

$$B(t) = \exp\left(\int_0^t r(u)du\right), \; B(0) = 1$$

The chain $\mathbf{X}$ determines the appreciation rate $\mu(t)$ and the volatility $\sigma(t)$ of the stock, respectively, as:

$$\mu(t) = \langle \boldsymbol{\mu}, \mathbf{X}(t) \rangle$$

and

$$\sigma(t) = \langle \boldsymbol{\sigma}, \mathbf{X}(t) \rangle$$

where $\boldsymbol{\mu} := (\mu_1, \mu_2, \ldots, \mu_N)' \in \mathbb{R}^n$, $\boldsymbol{\sigma} := (\sigma_1, \sigma_2, \ldots, \sigma_N)' \in \mathbb{R}^n$ and with $\mu_i > r_i$ and $\sigma_i > 0$, for each $i = 1, 2, \ldots, N$.

Let $w := \{w(t) | t \in \mathcal{T}\}$ denote a standard Brownian motion on $(\Omega, \mathcal{F}, \mathcal{P})$ with respect to the $\mathcal{P}$-augmentation of its own natural filtration. We suppose that w and $\mathbf{X}$ are stochastically independent. The evolution of the price process of the stock follows a Markovian regime-switching *GBM*:

$$dS(t) = \mu(t)S(t)dt + \sigma(t)S(t)dw(t),\ S(0) = s > 0$$

Now, we specify the information structure of our model. Let $\mathcal{F}^{\mathbf{X}}$ and $\mathcal{F}^{S}$ denote the right-continuous, complete filtrations generated by the values of the Markov chain and the stock price process, respectively. Write, for each $t \in \mathcal{T}$, $\mathcal{G}(t) := \mathcal{F}^{\mathbf{X}}(t) \vee \mathcal{F}^{S}(t)$, the enlarged σ-field generated by $\mathcal{F}^{\mathbf{X}}(t)$ and $\mathcal{F}^{S}(t)$.

In the sequel, we describe the evolution of the wealth process of an investor who allocates his/her wealth between the money market account and the stock. Let $\pi(t)$ denote the proportion of the total wealth invested in the stock at time $t \in \mathcal{T}$. Then, $1 - \pi(t)$ represents the proportion of the total wealth invested in the money market account at time t. We suppose that $\pi := \{\pi(t)\}_{t\in\mathcal{T}}$ is $\mathcal{G}$-progressively measurable and càdlàg (i.e. right continuous with left limit, RCLL). This means that the investor selects the proportion of wealth allocated to the stock based on information generated by the stock price process and the state of the economy.

We further assume that π is self-financing, (i.e. there is no income or consumption), and that

$$\int_0^T \pi^2(t)dt < \infty, \mathcal{P}\text{−a.s.}$$

Write $\mathcal{A}$ for the set of all such processes π We call A the set of admissible portfolio processes.

Let $V(t) := V^{\pi}(t)$ denote the total wealth of the portfolio π at time t. Then, the evolution of the wealth process over time is governed by:

$$dV(t) = V(t)\{[r(t) + \pi(t)(\mu(t) - r(t))]dt + \pi(t)\sigma(t)dw(t)\}, \quad V(0) = v > 0.$$

Our goal is to find a portfolio process π which minimizes the risk of the terminal wealth. Here, we use a particular form of convex risk measure introduced in [5] as a measure of risk.

7.2.2 *Risk Minimization*

In this section, we first describe the notion of convex risk measures. Then, we present the risk minimization problem of an investor with wealth process described in the last section and formulate the problem as a Markovian regime-switching version of a two-player, zero-sum, stochastic differential game.

The concept of a convex risk measure provides a generalization of a coherent risk measure. Suppose $\mathcal{S}$ denote the space of all lower-bounded, $g(T)$-measurable, random variables. A convex risk measure ρ is a functional $\rho : \mathcal{S} \to \mathbb{R}$ such that it satisfies the following three properties:

(1) If $X \in \mathcal{S}$ and $\beta \in \mathbb{R}$, then

$$\rho(X + \beta) = \rho(X) - \beta$$

(2) For any $X, Y \in \mathcal{S}$, if $X(\omega) \leq Y(\omega)$, for all $\omega \in \Omega$, then $\rho(X) \geq \rho(Y)$.
(3) For any $X, Y \in \mathcal{S}$ and $\lambda \in (0, 1)$,

$$\rho(\lambda X + (1 - \lambda)Y) \leq \lambda\rho(X) + (1 - \lambda)\rho(Y).$$

The first, second and third properties are the translation invariance, monotonicity and convexity, respectively.

Föllmer and Scheid [20] provide an elegant representation for convex risk measures. One can generate any convex risk measure from this representation by a suitable choice of a family of probability measures. Let $\mathcal{M}_a$ denote a family of probability measures $\mathcal{Q}$ which are absolutely continuous with respect to $\mathcal{P}$. Define a function $\eta : \mathcal{M}_a \to \mathbb{R}$ such that $\eta(\mathcal{Q}) < \infty$, for all $\mathcal{Q} \in \mathcal{M}_a$. Then, [21] provides the following representation of a convex risk measure $\rho(X)$ of $X \in \mathcal{S}$:

$$\rho(X) = \sup_{\mathcal{Q} \in \mathcal{M}_a} \{\mathbf{E}_{\mathcal{Q}}[-X] - \eta(\mathcal{Q})\},$$

for some family $\mathcal{M}_a$ and some function η.

Here, $\mathbf{E}_{\mathcal{Q}}[\cdot]$ represents expectation under $\mathcal{Q}$. The function $\eta(\cdot)$ is called a "penalty" function.

Following Mataramvura and Øksendal [17], we consider a particular form of convex risk measure. Let $\eta_0 : \mathbb{R} \to \mathbb{R}$ denote a real-valued function. Then, assume that the penalty function $\eta(\mathcal{Q})$ has the following form:

$$\eta(\mathcal{Q}) = \mathbf{E}\left[\eta_0\left(\frac{d\mathcal{Q}}{d\mathcal{P}}\right)\right].$$

Let $I(\mathcal{Q}, \mathcal{P})$ denote the relative entropy of a probability measure $\mathcal{Q}$ with respect to a prior probability $\mathcal{P}$. Then, when $\eta_0(x) = \alpha x \ln(x)$,

$$\eta(\mathcal{Q}) = \alpha I(\mathcal{Q}, \mathcal{P}).$$

In this case, the convex risk measure with the penalty function $\eta(Q)$ becomes the entropic risk measure with the risk tolerance level α. That is,

$$e_\alpha(X) = \sup_{Q \in M_a} \left\{\mathbf{E}_Q[-X] - \alpha I(Q, P)\right\}, \quad X \in S.$$

The entropic risk measure is a classical example of convex risk measure and is related to an exponential utility function as follows:

$$e_\alpha(X) = \alpha \ln \mathbf{E}\left[\exp\left(-\frac{1}{\alpha}X\right)\right]$$

In the sequel, we generate a family $\mathcal{M}_a$ of real-world probability measures, which are absolutely continuous with respect to $\mathcal{P}$, by a product of two density processes, one for the Brownian motion w and one for the Markov chain $\mathbf{X}$.

Define a Markovian regime-switching process $\theta(t)$ as:

$$\theta(t) = \langle \boldsymbol{\theta}, \mathbf{X}(t) \rangle,$$

where $\boldsymbol{\theta} := (\theta_1, \theta_2, \ldots, \theta_N)' \in \mathbb{R}^N$ with $\theta_{(N)} := \max_{1 \leq i \leq N} \theta_i < \infty$. Write Θ for the space of all such processes.

Consider a $\mathcal{G}$-adapted process $\Lambda^\theta := \left\{\Lambda^\theta(t)\right\}_{t \in \mathcal{T}}$:

$$\Lambda^\theta := \exp(-\int_0^t \theta(s)dw(s) - \frac{1}{2}\int_0^t \theta^2(s)ds).$$

Then, by Itô's differentiation rule,

$$d\Lambda^\theta = -\Lambda^\theta(t)\theta(t)dw(t), \Lambda^\theta(0) = 1.$$

So, Λ^θ is a $(\mathcal{G}, \mathcal{P})$-local-martingale.

Since the Novikov condition is satisfied here, Λ^θ is a $(\mathcal{G}, \mathcal{P})$ martingale, and $\mathbf{E}[\Lambda^\theta(T)] = 1$.

For each $i, j, k = 1, 2, \ldots, N$, define a real-valued, $\mathcal{F}^w$-adapted, stochastic process $c_{ij}^k(t) \geq 0$ such that for each $t \in \mathcal{T}$,

(1) $c_{ij}^k(t) \geq 0$, for $i \neq j$,
(2) $\sum_{i=1}^N c_{ij}^k(t) = 0$, so $c_{ii}^k(t) \leq 0$, for each $k = 1, 2, \ldots, N$.

Suppose $\mathbf{C}(t) := \left\{c_{ij}(t)\right\}_{i,j=1,2,\ldots,N}, t \in \mathcal{T}$, is a second family of generators, or rate matrices, of the chain X such that for each $i, j = 1, 2, \ldots, N$

$$c_{ij}(t) = \left\langle \mathbf{c}_{ij}(t), \mathbf{X}(t) \right\rangle,$$

where $\mathbf{c}_{ij}(t) := (c_{ij}^1(t), c_{ij}^2(t), \ldots, c_{ij}^N(t))' \in \mathbb{R}^N$.

So, all the components of $\mathbf{c}_{ij}(t)$ are adapted to the $\mathcal{F}^w$-filtration. The state of the chain $\mathbf{X}(t)$ selects which component $c_{ij}(t)$ is in force at time t. The dependence of $c_{ij}(t)$ on $\mathbf{X}(t)$ is obtained by taking the scalar product $\left\langle \mathbf{c}_{ij}(t), \mathbf{X}(t) \right\rangle$.

For each $k = 1, 2, \ldots, N$, write $\mathbf{C}^h(t) := \left[c_{ij}^k(t)\right]_{i,j=1,2,\ldots,N}$. Here, $\mathbf{C}^k(t)$ is a $\mathcal{F}^w(t)$-measurable, matrix-valued, random element. Then,

$$\mathbf{C}(t) = \sum_{k=1}^{N} c^k(t)\langle \mathbf{X}(t), \mathbf{e}_k \rangle.$$

We wish to introduce a new (real-world) probability measure under which $\mathbf{C}(t)$, $t \in \mathcal{T}$, is a family of generators of the chain $\mathbf{X}$. We follow the method in [22]. First, we define some notations. Let $\mathcal{C}$ denote the space of any such family $\mathbf{C}(t), t \in \mathcal{T}$. For any two matrices $\mathbf{A}(t)$, with $a_{ij}(t) \neq 0$, for any $t \in \mathcal{T}$ and $i, j = 1, 2, \ldots, N$, and $\mathbf{C}(t)$, write $\mathbf{D}(t) := \mathbf{C}(t)/\mathbf{A}(t)$ for the matrix defined by $\mathbf{D}(t) := \left[c_{ij}(t)/a_{ij}(t)\right], t \in \mathcal{T}$. Write $\mathbf{1} := (1, 1, \ldots, 1)' \in \mathbb{R}^N$ and $\mathbf{I}$ for the $(N \times N)$-identity matrix.

Define, for each $t \in \mathcal{T}$

$$\mathbf{N}(t) := \int_0^t (\mathbf{I} - \mathbf{diag}(\mathbf{X}(u-)))d\mathbf{X}(u),$$

Here, $\mathbf{N} := \{\mathbf{N}(t)\}_{t \in \mathcal{T}}$ is a vector of counting processes, where its component $N_i(t)$ counts the number of times the chain $\mathbf{X}$ jumps to the state $\mathbf{e}_i$ in the time interval $[0, t], i = 1, 2, \ldots, N$. Then, we cite the following result from [20] without proof.

Lemma 7.2.1 *For a given rate matrix* $\mathbf{A}(t)$, *write*

$$\mathbf{a}(t) := (a_{11}(t), \ldots, a_{ii}(t), \ldots, a_{NN}(t))',$$

and

$$\mathbf{A}_0(t) := \mathbf{A}(t) - \mathbf{diag}(\mathbf{a}(t)),$$

where $\mathbf{diag}(\mathbf{y})$ *is a diagonal matrix with the diagonal elements given by the vector* $\mathbf{y}$. *Define*

$$\widetilde{\mathbf{N}}(t) := \mathbf{N}(t) - \int_0^t \mathbf{A}_0(u)\mathbf{X}(u)du, t \in \mathcal{T},$$

Then, $\widetilde{\mathbf{N}} := \left\{\widetilde{\mathbf{N}}(t)\right\}_{t \in \mathcal{T}}$ is an $(\mathcal{F}^{\mathbf{X}}, \mathcal{P})$-martingale.

Consider a process $\Lambda^{\mathbf{C}} := \{\Lambda^{\mathbf{C}}(t)\}_{t \in \mathcal{T}}, \mathbf{C} \in \mathcal{C}$, defined by:

$$\Lambda^{\mathbf{C}}(t) = 1 + \int_0^t \Lambda^{\mathbf{C}}(u-)[\mathbf{D}_0(u)\mathbf{X}(u-) - \mathbf{1}]'(d\mathbf{N}(u) - \mathbf{A}_0(u)\mathbf{X}(u-)du).$$

Note that from Lemma 7.2.1, $\Lambda^{\mathbf{C}}$ is an $(\mathcal{F}^{\mathbf{X}}, \mathcal{P})$-martingale.

Define, for each $(\theta, \mathbf{C}) \in \Theta \times \mathcal{C}$, a $\mathcal{G}$-adapted process $\Lambda^{\theta,\mathbf{C}} := \{\Lambda^{\theta,\mathbf{C}}(t)\}_{t\in\mathcal{T}}$ as 其 the product of the two density processes Λ^{θ} and $\Lambda^{\mathbf{C}}$:

$$\Lambda^{\theta,\mathbf{C}} := \Lambda^{\theta}(t) \cdot \Lambda^{\mathbf{C}}(t).$$

Lemma 7.2.2 $\Lambda^{\theta,\mathbf{C}}$ *is a* $(\mathcal{G}, \mathcal{P})$*-martingale.*

The detail of the proof will be published in [18].

Define, for each $(\theta, \mathbf{C}) \in \Theta \times \mathcal{C}$, a real-world probability measure $\mathcal{Q}^{\theta,\mathbf{C}} \sim \mathcal{P}$ on as:

$$\frac{d\mathcal{Q}^{\theta,\mathbf{C}}}{d\mathcal{P}}\Big|_{\mathcal{G}(T)} := \Lambda^{\theta,\mathbf{C}}(T). \tag{7.2.1}$$

Then, we generate a family of $\mathcal{M}_a$ of real-world probability measures as follows:

$$\mathcal{M}_a := \mathcal{M}_a(\Theta, \mathcal{C}) = \{\mathcal{Q}^{\theta,\mathbf{C}} | (\theta, \mathbf{C}) \in \Theta \times \mathcal{C}\}.$$

The following result is from [20]. We cite it in the following lemma without giving the proof.

Lemma 7.2.3 *Suppose* $\mathcal{Q}^{\theta,\mathbf{C}}$ *is defined by (1), for each* $(\theta, \mathbf{C}) \in \Theta \times \mathcal{C}$*. Let*

$$\widetilde{\mathbf{N}}^{\mathbf{C}}(t) := \mathbf{N}(t) - \int_0^t \mathbf{C}_0(u)\mathbf{X}(u)du, t \in \mathcal{T}, \mathbf{C} \in \mathcal{C},$$

Then, $\widetilde{\mathbf{N}}^{\mathbf{C}} := \left\{\widetilde{\mathbf{N}}^{\mathbf{C}}(t)\right\}_{t\in\mathcal{T}}$ *is an* $(\mathcal{F}^{\mathbf{X}}, \mathcal{Q}^{\theta,\mathbf{C}})$*-martingale.*

Theorem 7.2.4 *For each* $(\theta, \mathbf{C}) \in \Theta \times \mathcal{C}$, $\mathbf{X}$ *is a Markov chain with a family of generators* $\mathbf{C}(t), t \in \mathcal{T}$ *under* $\mathcal{Q}^{\theta,\mathbf{C}}$.

According to [17], we define a vector process $\mathbf{Z} := \{\mathbf{Z}(t)\}_{t\in\mathcal{T}}$ by

$$\begin{aligned}
d\mathbf{Z}(t) &= (dZ_0(t), dZ_1(t), dZ_2(t), dZ_3(t), d\mathbf{Z}_4(t))' \\
&= \left(dZ_0(t), dZ_1^{\pi}(t), dZ_2^{\theta}(t), dZ_3^{\mathbf{C}}(t), d\mathbf{Z}_4(t)\right)' \\
&= \left(dZ_0(t), dV^{\pi}(t), d\Lambda^{\theta}(t), d\Lambda^{\mathbf{C}}(t), d\mathbf{X}(t)\right)', \\
\mathbf{Z}(0) &= \mathbf{z} = (s, z_1, z_2, z_3, \mathbf{z}_4)'.
\end{aligned}$$

where under $\mathbb{R}$,

$$
\begin{aligned}
dZ_0(t) &= dt,\\
Z_0(0) &= s \in \mathcal{T},\\
dZ_1(t) &= Z_1(t)\{[r(t) + (\mu(t) - r(t))\pi(t))]dt + \sigma(t)\pi(t)dw(t)\},\\
Z_1(0) &= z_1 > 0,\\
dZ_2(t) &= -\theta(t)Z_2(t)dw(t),\\
Z_2(0) &= z_2 > 0,\\
dZ_3(t) &= Z_3(t-)(\mathbf{D}_0(t)\mathbf{X}(t-) - \mathbf{1})'(d\mathbf{N}(t) - \mathbf{A}_0(t)\mathbf{X}(t-)dt),\\
Z_3(0) &= z_3 > 0,\\
d\mathbf{Z}_4(t) &= \mathbf{A}(t)\mathbf{Z}_4(t-)dt + d\mathbf{M}(t),\\
\mathbf{Z}_4(0) &= \mathbf{z}_4.
\end{aligned}
$$

Conditional on $\mathbf{Z}(0) = \mathbf{z}$, the penalty function is given by:

$$\eta^Z(\mathcal{Q}) := \mathbf{E}^z[\eta_0(\frac{d\mathcal{Q}}{d\mathcal{P}})],$$

where $\mathbf{E}^z[\cdot]$ represents expectation under $\mathcal{P}$ given that the initial value $\mathbf{Z}(0) = \mathbf{z}$.

So, for each $\mathcal{F}^W$-adapted process $(\theta, \mathbf{C}) \in \Theta \times \mathcal{C}$, we define the induced penalty function $\bar{\eta}^{\mathbf{Z}}(\theta, \mathbf{C})$ as:

$$\bar{\eta}^{\mathbf{Z}}(\theta, \mathbf{C}) = \eta^{\mathbf{Z}}(\mathcal{Q}^{\theta,\mathbf{C}}) = \mathbf{E}^{\mathbf{Z}}[\eta_0(\ln(Z_2^\theta(T)) + \ln(Z_3^{\mathbf{C}}(T)))].$$

Now, conditional on $\mathbf{Z}(0) = \mathbf{z}$, the risk-minimizing problem is then to find a portfolio process $\pi \in \mathcal{A}$ in order to minimize the following conditional version of the convex risk measure associated with $\Theta \times \mathcal{C}$:

$$\sup_{(\theta,\mathbf{C})\in\Theta\times\mathcal{C}} \left\{\mathbf{E}^{\mathbf{Z}}_{(\theta,\mathbf{C})}[-Z_1^\pi(T)] - \bar{\eta}^{\mathbf{Z}}(\theta, \mathbf{C})\right\},$$

where $\mathbf{E}^{\mathbf{z}}_{(\theta,\mathbf{C})}[\cdot]$ denotes expectation under $\mathcal{Q}^{\theta,\mathbf{C}}$ given that $\mathbf{Z}(0) = \mathbf{z}$.

This is equivalent to the following zero-sum stochastic differential game between the investor and the market:

$$
\begin{aligned}
\Phi(\mathbf{z}) &= \inf_{\pi\in\mathcal{A}}\left(\sup_{(\theta,\mathbf{C})\in\Theta\times\mathcal{C}} \left\{\mathbf{E}^{\mathbf{Z}}_{(\theta,\mathbf{C})}[-Z_1^\pi(T)] - \bar{\eta}^{\mathbf{Z}}(\theta, \mathbf{C})\right\}\right)\\
&= \mathbf{E}^{\mathbf{Z}}_{(\theta^*,\mathbf{C}^*)}[-Z_1^{\pi*}(T)] - \bar{\eta}^{\mathbf{Z}}(\theta^*, \mathbf{C}^*).
\end{aligned}
$$

To solve the game, we need to find the value function $\Phi(\mathbf{z})$ and the optimal strategies $\pi^* \in \mathcal{A}$, $(\theta^*, \mathbf{C}^*) \in \Theta \times \mathcal{C}$ of the investor and the market, respectively.

7.2.3 Solution to the Risk-Minimizing Problem

Following the plan in [17], we restrict ourselves to consider only Markovian controls for the risk-minimizing problem. Suppose $\vartheta := (0, T) \times (0, \infty) \times (0, \infty) \times (0, \infty)$ representing our solvency region. Let K_1 denote the set such that $\pi(t) \in K_1$. To restrict ourselves to Markovian controls, we assume that

$$\pi(t) := \bar{\pi}(\mathbf{Z}(t)).$$

Here, we do not distinguish between π and $\bar{\pi}$. So, we can simply identify the control process with deterministic function $\pi(\mathbf{z}), \mathbf{z} \in \vartheta \times \mathcal{E}$. This is called a feedback control.

We also suppose that the components of $\theta(t)$ and $\mathbf{c}_{ij}(t)$ are Markovian in w and that the dependence of $\theta(t)$ and $\mathbf{c}_{ij}(t)$ on $\mathbf{X}(t)$ are modeled by scalar products. In this case, $(\theta(t), \mathbf{C}(t))$ is also Markovian with respect to $\mathcal{G}$. So, the control processes $(\theta(t), \mathbf{C}(t), \pi(t))$ are Markovian. They are also feedback control processes since they depend on the current value of the state process $\mathbf{Z}(t)$.

Consider a process $Y := \{Y(t)\}_{t \in \mathcal{T}}$ defined by:

$$dY(t) = (\mathbf{D}_0(t)\mathbf{X}(t-) - \mathbf{1})' d\mathbf{N}(t).$$

From $d\mathbf{N}(t) = (\mathbf{I} - \mathbf{diag}(\mathbf{X}(t-)))d\mathbf{X}(t)$, so

$$dY(t) = (\mathbf{D}_0(t)\mathbf{X}(t-) - \mathbf{1})'(\mathbf{I} - \mathbf{diag}(\mathbf{X}(t-)))d\mathbf{X}(t).$$

Let $\Delta Y(t)$ denote the jump of the process Y at time t. Then

$$\begin{aligned}\Delta Y(t) &:= Y(t) - Y(t-) \\ &= (\mathbf{D}_0(t)\mathbf{X}(t-) - \mathbf{1})'(\mathbf{I} - \mathbf{diag}(\mathbf{X}(t-)))\Delta\mathbf{X}(t) \\ &= (\mathbf{D}_0(t)\mathbf{X}(t-) - \mathbf{1})'(\mathbf{I} - \mathbf{diag}(\mathbf{X}(t-)))(\mathbf{X}(t) - \mathbf{X}(t-)).\end{aligned}$$

By some algebra,

$$\Delta Y(t) = \sum_{i,j=1}^{N} (d_{ji} - 1)\langle \mathbf{X}(t), \mathbf{e}_j \rangle \langle \mathbf{X}(t-), \mathbf{e}_i \rangle.$$

Define, for each $i = 1, 2, \ldots, N$, the set

$$y_i := \{d_{1i} - 1, d_{2i} - 1, \ldots, d_{Ni} - 1\}.$$

Consider a random set $y(\mathbf{X}(t))$ defined by

$$y(\mathbf{X}(t)) = \sum_{i=1}^{N} y_i \langle \mathbf{X}(t), \mathbf{e}_i \rangle, t \in \mathcal{T}.$$

Let $y := \cup_{i=1}^{N} y_i$, then

$$y = \{d_{ji} - 1 | i,j = 1,2,\ldots,N\}.$$

Clearly, $y(\mathbf{X}(t)) \subset y, t \in \mathcal{T}$.

Given $\mathbf{X}(t-) = \mathbf{e}_i (i = 1,2,\ldots,N)$, y_i represents the set of all possible values of the jump $\Delta Y(t)$ at time t. The random set $y(\mathbf{X}(t))$ represents the set of possible values of the jump $\Delta Y(t)$ conditional on the value of $\mathbf{X}(t)$.

Suppose γ denotes the random measure which selects the jump times and sizes of the process Y. Let $\delta_a(\cdot)$ denote the Dirac measure, or the point mass, at $a \in \mathbb{R}$. Then, for each $K \in y$, the random measure is:

$$\begin{aligned}\gamma(t,K;\omega) &= \sum_{0<u\leq t} I_{\{\Delta Y(u) \in K, \Delta Y(u) \neq 0\}} \\ &= \sum_{0<u\leq t} I_{\{\Delta Y(u) \neq 0\}} \delta_{(u,\Delta Y(u))}((0,t] \times K).\end{aligned}$$

To simplify the notation, we suppress the subscript ω and write $\gamma(t,K) := \gamma(t,K;\omega)$.

$\gamma(dt,dy)$ denote the differential form of $\gamma(t,K)$. Define, for each $i = 1,2,\ldots,N$, a probability mass function $n_i(\cdot,t)$ on y_i as:

$$n_i(d_{ji} - 1, t) = a_{ji}(t).$$

Then, the predictable compensator of $\gamma(dt,dy)$ is:

$$\nu_{X(t-)}(dt,dy) = \sum_{i=1}^{N} n_i(dy,t-)\langle \mathbf{X}(t-), \mathbf{e}_i \rangle dt.$$

Write $\tilde{\gamma}(dt,dy)$ for the compensated version of the random measure $\gamma(dt,dy)$. That is,

$$\tilde{\gamma}(dt,dy) := \gamma(dt,dy) - \nu_{\mathbf{X}(t-)}(dt,dy).$$

Let $\mathcal{H}$ denote the space of functions $h(\cdot,\cdot,\cdot,\cdot,\cdot) : \mathcal{T} \times (\Re^+)^3 \times \mathcal{E} \to \mathbb{R}$ such that for each $\mathbf{x} \in \mathcal{E}$, $h(\cdot,\cdot,\cdot,\cdot,\mathbf{x})$ is $\mathcal{C}^{1,2,1}(\mathcal{T} \times (\mathbb{R}^+)^3)$. Write

$$\mathbf{H}(s,z_1,z_2,z_3) := (h(s,z_1,z_2,z_3,\mathbf{e}_1),\ldots,h(s,z_1,z_2,z_3,\mathbf{e}_N))' \in \mathbb{R}^N.$$

Define the Markovian regime-switching generator L $\mathcal{L}^{\theta,\mathbf{C},\pi}$ acting on a function $h \in \mathcal{H}$ for a Markov process $\{\mathbf{Z}^{\theta,\mathbf{C},\pi}(t)\}_{t\in T}$ as:

$$\begin{aligned}
&\mathcal{L}^{\theta,\mathbf{C},\pi}[h(s,z_1,z_2,z_3,\mathbf{z}_4)] \\
&= \frac{\partial h}{\partial s} + z_1[r(s) + (\mu(s) - r(s))\pi(\mathbf{z})]\frac{\partial h}{\partial z_1} + \frac{1}{2}\theta^2(s)z_2^2\frac{\partial^2 h}{\partial z_2^2} + \frac{1}{2}z_1^2\pi^2(\mathbf{z})\partial^2(t)\frac{\partial^2 h}{\partial z_1^2} \\
&\quad - \theta(s)\pi(\mathbf{z})z_1z_2\sigma(s)\frac{\partial^2 h}{\partial z_1 \partial z_2} + \int_{y(\mathbf{x})} (h(s,z_1,z_2,z_3(1+y),\mathbf{z}_4) \\
&\quad - h(s,z_1,z_2,z_3,\mathbf{z}_4) - \frac{\partial h}{\partial z_3} z_3 y)\nu_{\mathbf{x}}(ds,dy) + \langle \mathbf{H}(s,z_1,z_2,z_3), \mathbf{A}(s)\mathbf{x}\rangle.
\end{aligned}$$

Then, we need the following lemma for the development of a verification theorem of the HJB solution to the stochastic differential game. This lemma can be proof by using the generalized Itô's formula and conditioning on $\mathbf{Z}(0) = \mathbf{z}$ under $\mathcal{P}$.

Lemma 7.2.5 *Let $\tau < \infty$ be a stopping time. Assume further that $h(\mathbf{Z}(t))$ and $\mathcal{L}^{\theta,\mathbf{C},\pi}[h(\mathbf{Z}(t))]$ are bounded on $t \in [0,\tau]$.*

Then,

$$\mathbf{E}[h(\mathbf{Z}(\tau))|\mathbf{Z}(0) = \mathbf{z}] = h(\mathbf{z}) + \mathbf{E}\left[\int_0^\tau \mathcal{L}^{\theta,\mathbf{C},\pi}[h(\mathbf{Z}(t))]dt|\mathbf{Z}(0) = \mathbf{z}\right].$$

With the components of the controls $\theta(t)$ and $\mathbf{c}_{ij}(t)$ being Markovian in w and the dependence of them on the chain $\mathbf{X}(t)$ specified by the scalar products, the dynamic programming argument works well. We now describe the solution of the stochastic differential game between the investor and the market by the following verification theorem.

Theorem 7.2.1 *Let $\bar{\vartheta}$ denote the closure of ϑ. Suppose there exists a function ϕ such that for each $\mathbf{x} \in \mathcal{E}$, $\phi(\cdot,\cdot,\cdot,\cdot,\mathbf{x}) \in \mathcal{C}^2(\vartheta) \cap \mathcal{C}(\bar{\vartheta})$ and a Markovian control $(\widehat{\theta}(t), \widehat{\mathbf{C}}(t), \widehat{\pi}(t)) \in \Theta \times \mathcal{C} \times \mathcal{A}$, such that:*

(1) $\mathcal{L}^{\theta,\mathbf{C},\widehat{\pi}}[\phi(s,z_1,z_2,z_3,\mathbf{x})] \leq 0$, *for all* $(\theta,\mathbf{C}) \in \Theta \times \mathcal{C}$ *and,*

(2) $\mathcal{L}^{\widehat{\theta},\widehat{\mathbf{C}},\pi}[\phi(s,z_1,z_2,z_3,\mathbf{x})] \geq 0$, *for all* $\pi \in \mathcal{A}$ *and* $(s,z_1,z_2,z_3,\mathbf{x}) \in \vartheta \times \mathcal{E}$,

(3) $\mathcal{L}^{\widehat{\theta},\widehat{\mathbf{C}},\widehat{\pi}}[\phi(s,z_1,z_2,z_3,\mathbf{x})] = 0$, *for all* $(s,z_1,z_2,z_3,\mathbf{x}) \in \vartheta \times \mathcal{E}$,

(4) *for all* $(\theta,\mathbf{C},\pi) \in \Theta \times \mathcal{C} \times \mathcal{A}$,

$$\begin{aligned}
\lim_{t\to T-} \phi(t, Z_1^\pi(t), Z_2^\theta(t), Z_3^C(t), \mathbf{X}(t)) &= -Z_2^\theta(T)Z_3^C(T)Z_1^\pi(T) \\
&\quad - \eta_0(\ln(Z_2^\theta(T)) + \ln(Z_3^C(T))),
\end{aligned}$$

(5) let $\mathcal{K}$ *denote the set of stopping times* $\tau \leq T$. *The family* $\{\phi(\mathbf{Z}^{\theta,\mathbf{C},\pi}(\tau))\}_{\tau} \in \mathcal{K}$ *is uniformly integrable.*

Write, for each $\mathbf{z} \in \vartheta \times \mathcal{E}$ 和 $(\theta, \mathbf{C}, \pi) \in \Theta \times \mathcal{C} \times \mathcal{A}$,

$$J^{\theta,\mathbf{C},\pi}(\mathbf{z}) := E^{\mathbf{z}}_{(\theta,\mathbf{C})}\{-Z_2^{\theta}(T-s)Z_3^{C}(T-s)Z_1^{\pi}(T-s) \\ - \eta_0(\ln(Z_2^{\theta}(T-s)) + \ln(Z_3^{C}(T-s))\},$$

Then

$$\begin{aligned}\phi(\mathbf{z}) &= \Phi(\mathbf{z}) \\ &= \inf_{\pi\in\mathcal{A}}(\sup_{(\theta,\mathbf{C})\in\Theta\times\mathcal{C}} J^{\theta,\mathbf{C},\pi}(\mathbf{z})) \\ &= \sup_{(\theta,\mathbf{C})\in\Theta\times\mathcal{C}}(\inf_{\pi\in\mathcal{A}} J^{\theta,\mathbf{C},\pi}(\mathbf{z})) \\ &= \inf_{\pi\in\mathcal{A}} J^{\widehat{\theta},\widehat{\mathbf{C}},\pi}(\mathbf{z}) = \sup_{(\theta,\mathbf{C})\in\Theta\times\mathcal{C}} J^{\theta,\mathbf{C},\widehat{\pi}}(z) \\ &= J^{\widehat{\theta},\widehat{\mathbf{C}},\widehat{\pi}}(\mathbf{z}).\end{aligned}$$

and $\left(\widehat{\theta}, \widehat{\mathbf{C}}, \widehat{\pi}\right)$ is an optimal Markovian control.

The proof is adapted from the proof of Theorem 3.2 in [17] and uses Lemma 7.2.5 here.

7.3 Option Pricing Based on Game Theory

In finance, an option is a contract which gives the buyer (the owner or holder) the right, but not the obligation, to buy or sell an underlying asset or instrument at a specified strike price on or before a specified date, depending on the form of the option. Option pricing is the core issue of option trading because it is the only variable changes with the market supply and demand in option contract, and directly affect the profit and loss situation of buyers and sellers. It occured the first paper about option price in 1900. Since then, a variety of empirical formulas or metered pricing models have been available, but it is difficult to get generally recognized because of various limitations. Since the 1970s, the research of option pricing theory had made a breakthrough along with the development of options market.

A very important issue about options which plagued investors is how to determine their values in the formation process of the international derivatives markets. The application of computer and advanced communication technology made complex option pricing formula possible. In the past 20 years, the investors transformed this abstract numerical formula into a great deal of wealth through Black-Scholes option pricing model.

7.3.1 Black-Scholes-Meton Market Model

We consider a continuous-time financial model with two primitive assets, that are tradable continuously on a finite time horizon $\mathcal{T} := [0, T]$, where $T \in (0, \infty)$. Then, we fix a complete probability space $(\Omega, \mathcal{F}, \mathcal{P})$, where $\mathcal{P}$ is a real-world probability measure. we use boldface letters to denote vectors or matrices. Let $\mathbf{X} := \{\mathbf{X}(t)\}_{t \in \mathcal{T}}$ denote an observable, continuous-time and finite-state Markov chain on $(\Omega, \mathcal{F}, \mathcal{P})$ with a finite state space $\mathcal{S} := \{\mathbf{s}_1, \mathbf{s}_2, \ldots, \mathbf{s}_N\} \subset \mathbb{R}^N$. We identify the state space of the chain $\mathbf{X}$ to be a finite set of unit vectors $\mathcal{S} := \{\mathbf{s}_1, \mathbf{s}_2, \ldots, \mathbf{s}_N\} \subset \mathbb{R}^N$, where $\mathbf{e}_i \in \mathbb{R}^N$ and the jth component of $\mathbf{e}_i$ is the Kronecker delta δ_{ij}, for each $i, j = 1, 2, \ldots, N$, y' represents the transpose of a vector or a matrix y. The set $\mathcal{E}$ is called the canonical state space of $\mathbf{X}$. Let $\Pi = [a_{ij}]_{i,j=1,2,\ldots,N}$ denote the rate matrix for the chain $\mathbf{X}$. Then, Elliott et al. (1994) [19] provided the following semi-martingale decomposition for $\mathbf{X}$:

$$\mathbf{X}(t) = \mathbf{X}(0) + \int_0^t \Pi \mathbf{X}(u) du + \mathbf{M}(t), \tag{7.3.1}$$

where $\{\mathbf{M}(t)\}_{t \in \mathcal{T}}$ is an $\mathbb{R}^N$-valued martingale with respect to the $\mathcal{P}$ augmentation of the natural filtration generated by $\mathbf{X}$. The semimartingale decomposition describes the evolution of the chain.

Let $r(t)$ denote the instantaneous market interest rate of the money market account B at time t. We suppose that

$$r(t) = \langle \mathbf{r}, \mathbf{X}(t) \rangle,$$

where $\mathbf{r} = (r_1, r_2, \ldots, r'_N \in \mathbb{R}^N$ with, $r_i > 0$, for each $i = 1, 2, \ldots, N$, $\langle \cdot, \cdot \rangle$ denotes an inner product.

The price dynamics of the money market account:

$$B(t) = \exp(\int_0^t r(u) du), \ t \in \mathcal{T}, \ B(0) = 1. \tag{7.3.2}$$

Let $\mu(t)$ and $\sigma(t)$ denote the appreciation rate and the volatility of the stock S at time $t \in \mathcal{T}$, respectively, which are assumed to be governed by:

$$\mu(t) = \langle \boldsymbol{\mu}, \mathbf{X}(t) \rangle, \ \sigma(t) = \langle \boldsymbol{\sigma}, \mathbf{X}(t) \rangle \tag{7.3.3}$$

where $\boldsymbol{\mu} = (\mu_1, \mu_2, \ldots, \mu_N)' \in \mathbb{R}^N$ and $\boldsymbol{\sigma} = (\sigma_1, \sigma_2, \ldots, \sigma_N)' \in \mathbb{R}^N$ with $\mu_i > r_i$ and $\sigma_i > 0$, for each $i = 1, 2, \ldots, N$.

Let $w := \{w(t) | t \in \mathcal{T}\}$ denote a standard Brownian motion on $(\Omega, \mathcal{F}, \mathcal{P})$ with respect to the $\mathcal{P}$ augmentation. We suppose that w and $\mathbf{X}$ are independent. The price dynamics of the stock $\{S(t) | t \in \mathcal{T}\}$ are assumed to be governed by the following Markovian regime-switching geometric Brownian motion (GBM):

$$dS(t) = \mu(t)S(t)dt + \sigma(t)S(t)dw(t),\ S(0) = s > 0. \tag{7.3.4}$$

Let $Y(t) = \ln[S(t)/S(0)]$, which is the log return from the risky asset Sover the time interval $[0, t]$, for each $t \in \mathcal{T}$. Then,

$$dY(t) = (\mu(t) - \frac{1}{2}\sigma^2(t))dt + \sigma(t)dw(t). \tag{7.3.5}$$

We then specify the information structure of our model. Let $\mathcal{F}^{\mathbf{X}} = \{\mathcal{F}^{\mathbf{X}}(t) | t \in \mathcal{T}\}$ and $\mathcal{F}^Y = \{\mathcal{F}^Y(t) | t \in \mathcal{T}\}$ denote right continuous, complete filtrations generated by the processes $\mathbf{X}$ and Y, respectively. For each $t \in \mathcal{T}$, let $\mathcal{G}(t) = \mathcal{F}^{\mathbf{X}}(t) \vee \mathcal{F}^Y(t)$, an enlarged filtration generated by both $\mathbf{X}$ and Y. Write $\mathcal{G} = \{\mathcal{G}(t) | t \in \mathcal{T}\}$.

Define, for each $t \in \mathcal{T}$,

$$\theta(t) = \langle \boldsymbol{\theta}, \mathbf{X}(t) \rangle, \tag{7.3.6}$$

where $\boldsymbol{\theta} = (\theta_1, \theta_2, \ldots, \theta_N) \in \mathbb{R}^N$ with $\theta_{(N)} = \max_{1 \le i \le N} \theta_i < \infty$.

Let Θ denote the space of all such processes $\theta = \{\theta(t) | t \in \mathcal{T}\}$. We define a real-valued $\mathcal{G}$-adapted process associated with $\theta \in \Theta$ on $(\Omega, \mathcal{F}, \mathcal{P})$ as below:

$$\begin{aligned} \Lambda^\theta(t) &= \exp\left(- \int_0^t \theta(u)dw(u) - \frac{1}{2}\int_0^t \theta^2(u)du \right) \\ &= \exp\left(- \sum_{i=1}^N \theta_i^2 \int_0^t \langle \mathbf{X}(t), \mathbf{e}_i \rangle dw(u) - \frac{1}{2}\sum_{i=1}^N \theta_i^2 \int_0^t \langle \mathbf{X}(t), \mathbf{e}_i \rangle du \right), \end{aligned} \tag{7.3.7}$$

where $\int_0^t \langle \mathbf{X}(t), \mathrm{e}_i \rangle dW(u)$ represents the level integral of $\mathbf{X}$ with respect to w, and $\int_0^t \langle \mathbf{X}(t), \mathbf{e}_i \rangle du$ is the occupation time of $\mathbf{X}$in State i over the time duration $[0, t]$.

Then, by Itô's rule,

$$d\Lambda^\theta(t) = -\Lambda^\theta(t)\theta(t)dw(t),\ t \in \mathcal{T},\ \Lambda^\theta(0) = 1. \tag{7.3.8}$$

Here, Λ^θ is a local-martingale with respect to $(\mathcal{G}, \mathcal{P})$.

$$\text{Note that } \mathrm{E}\left[\exp(\frac{1}{2}\int_0^t \theta^2(t)dt)\right] < \exp(\frac{1}{2}\theta_N^2 T) < \infty, \tag{7.3.9}$$

where $\mathbf{E}[\cdot]$ denotes expectation under $\mathcal{P}$. From (7.3.9), Novikov's condition is satisfied. Hence, Λ^θ is a $(\mathcal{G}, \mathcal{P})$-martingale. This implies that $\mathbf{E}\left[\Lambda^\theta(t) = 1\right]$, $t \in \mathcal{T}$.

Then, for each $\theta \in \Theta$, we define a probability measure $\mathcal{P}^\theta$ equivalent to $\mathcal{P}$ on $\mathcal{G}(T)$ as follows:

$$\left.\frac{d\mathcal{P}^\theta}{d\mathcal{P}}\right|_{\mathcal{G}(T)} := \Lambda^\theta(T). \tag{7.3.10}$$

By Girsanov's theorem,

$$w^\theta := w(t) + \int_0^t \theta(u)du,\ t \in \mathcal{T}, \tag{7.3.11}$$

is a standard Brownian motion with respect to the enlarged filtration $\mathcal{G}$ under $\mathcal{P}$.

Then, under $\mathcal{P}^\theta$, the price dynamics of the risky asset S are governed by:

$$dS(t) = [\mu(t) - \theta(t)\sigma(t)]S(t)dt + \sigma(t)S(t)dw^\theta(t). \tag{7.3.12}$$

Using the above Girsanov-type transformation, we generate a family of probability measures $P_\Theta := \{\mathcal{P}^\theta\}_{\theta\in\Theta}$ equivalent to the reference probability $\mathcal{P}$ associated with the index set Θ. In other words, the family of probability measures P_Θ is parameterized by the space Θ of processes ϑ.

In the sequel, we present the stochastic differential game in the Markovian regime-switching Black–Scholes–Merton economy.

Here, the market selects a probability measure, or a generalized "scenario" in the context of coherent risk measures from the family P_Θ. This is equivalent to selecting a process $\theta \in \Theta$. So, Θ is the set of admissible controls of the market. On the other hand, the representative agent selects a portfolio that maximizes his/her expected utility of the terminal wealth.

We describe in some detail the portfolio process in the sequel. For each $t \in \mathcal{T}$, let $\pi(t)$ denote the proportion of wealth invested in the stock S at time t. We suppose that the portfolio process $\pi = \{\pi(t)|t \in \mathcal{T}\}$ is $\mathcal{G}$-progressively measurable and is self-financing. Define $V^\pi = \{V^\pi(t)|t \in \mathcal{T}\}$ as the wealth process corresponding to the portfolio process π. Then, under $\mathcal{P}$, the evolution of the wealth process is governed by the following stochastic differential equation (SDE):

$$dV^\pi(t) = V^\pi(t)[r(t) + (\mu(t) - r(t))\pi(t)]dt + V^\pi(t)\sigma(t)\pi(t)dw(t), \tag{7.3.13}$$

where $V^\pi(0) = v > 0$.

Under $\mathcal{P}^{\theta}$, the wealth process becomes:

$$\begin{aligned} dV^{\pi}(t) = V^{\pi}(t)[r(t) + (\mu(t) - r(t) - \theta(t)\sigma(t))\pi(t)]dt \\ + V^{\pi}(t)\sigma(t)\pi(t)dw^{\theta}(t). \end{aligned} \tag{7.3.14}$$

Let $\mathcal{A}$ denote the set of portfolio processes π such that

$$\int_0^T \left[r(t) + |\mu(t) - r(t)||\pi(t)| + \sigma^2(t)\pi^2(t)\right]dt < \infty, \quad \mathcal{P}\text{–a.s.} \tag{7.3.15}$$

We call $\mathcal{A}$ the set of admissible portfolios for the representative agent.

We suppose that the representative agent has the following power utility:

$$U(v) = \frac{v^{1-\zeta}}{1-\zeta}, \ \zeta \in (0,1), \ v \in [0,\infty), \tag{7.3.16}$$

Here, ζ is the risk aversion parameter of the power utility and the relative risk aversion of the representative agent is $1-\zeta$. So, the degree of risk aversion increases as ζ decreases.

Given a generalized "scenario" $\mathcal{P}^{\theta} \in P_{\Theta}$ chosen by the market, the representative agent chooses a portfolio process π so as to maximize the expected utility of terminal wealth with respect to the measure $\mathcal{P}^{\theta}$. Then, the response of the market to this choice is to select the generalized scenario $\mathcal{P}^{\theta}$ that minimizes the maximal expected utility. This situation can be formulated as a zero-sum stochastic differential game between the representative agent and the market. To ensure that a representative agent with an increasing and strictly concave utility function can be constructed, one may assume that the equilibrium allocation is Pareto optimal or Pareto efficient.

Now, we define a vector process $\mathbf{Z} := \{\mathbf{Z}(t)|t \in \mathcal{T}\}$:

$$\begin{aligned} d\mathbf{Z}(t) &= (dZ_0(t), dZ_1(t), dZ_2(t))' = (dt, d\Lambda^{\theta}(t), dV^{\pi}(t))', \\ \mathbf{Z}(0) &= \mathbf{z} = (u, z_1, z_2)' \in \mathcal{T} \times \mathbb{R}^2. \end{aligned}$$

In fact, $\mathbf{Z}(t) := \mathbf{Z}^{\theta,\pi}(t)$; that is $\mathbf{Z}(t)$ depends on θ and π. However, for notational simplicity, we suppress the subscripts θ and π.

Then, conditional on $\mathbf{Z}(0) = \mathbf{z}$ and $\mathbf{X}(0) = \mathbf{x} \in \mathcal{E}$ the stochastic differential game can be solved by finding the value function $\Phi(\mathbf{z}, \mathbf{x})$, the optimal strategies $\hat{\theta} \in \Theta$ and $\hat{\pi} \in \mathcal{A}$ such that

$$\begin{aligned} \Phi(\mathbf{z}, \mathbf{x}) &= \mathbf{E}^{\hat{\theta}}\left[U(V^{\hat{\pi}}(T))|(\mathbf{Z}(0), \mathbf{X}(0)) = (\mathbf{z}, \mathbf{x})\right] \\ &= \inf_{\theta \in \Theta}\left(\sup_{\pi \in \mathcal{A}} \mathbf{E}^{\theta}[U(V^{\pi}(T))|(\mathbf{Z}(0), \mathbf{X}(0)) = (\mathbf{z}, \mathbf{x})]\right). \end{aligned} \tag{7.3.17}$$

Here $\mathbf{E}^{\theta}[\cdot]$ represents expectation under $\mathcal{P}^{\theta}$ and $U(\cdot)$ is the power utility function of the representative agent defined in (7.3.16).

7.3.2 A Pricing Measure for Black-Scholes-Merton Market

Traditionally, the asset pricing theory has a closed connection to the theory of optimal portfolio and consumption decisions via the relationship between state prices and the marginal rates of substitution at optimality. This connection might be tracked back to the foundation of the price theory in economics where there is a closed link between the theory of optimal economic resources and the determination of prices [23, 24], Typically, in an equilibrium approach of asset pricing, the pricing problem is formulated as the optimal portfolio and consumption problem of a representative agent in a continuous-time version of the Lucas (1978) [25] exchange economy. In equilibrium, an Euler condition for the representative agent's optimal choices is derived and it forms a restriction on the security prices. The Euler condition involves the marginal rates of substitution, which are related to the state prices or equivalent martingale measures.

In this section, we first derive the solution to the zero-sum, two person stochastic differential game described in the last section. Consequently, the state prices or equivalent martingale measures are determined by the equilibrium state of the game, which involves not only the optimal portfolio choice of a representative agent, but also the optimal choice of a generalized "scenario" by the market.

First, we note that under $\mathcal{P}$ the evolutions of the components of the vector process $\mathbf{Z} = \{\mathbf{Z}(t)\}_{t\in\Gamma}$ are governed by:

$$\begin{aligned} dZ_0(t) &= dt, \\ dZ_1(t) &= -\theta(t)Z_1(t)dw(t), \\ dZ_2(t) &= Z_2(t)[r(t) + (\mu(t) - r(t))\pi(t)]dW(t) + Z_2(t)\sigma(t)\pi(t)dw(t), \end{aligned} \tag{7.3.18}$$

where $\mathbf{Z}(0) = \mathbf{z} = (u, z_1, z_2)$.

By the Bayes' rule,

$$\begin{aligned} \Phi(\mathbf{z}, \mathbf{x}) &= \mathbf{E}[\Lambda^{\hat{\theta}}(T)U(V^{\hat{\pi}}(T))|(\mathbf{Z}(0), \mathbf{X}(0)) = (\mathbf{z}, \mathbf{x})] \\ &= \inf_{\theta\in\Theta}\left(\sup_{\pi\in\mathcal{A}} \mathbf{E}[\Lambda^{\theta}(T)U(V^{\pi}(T))|(\mathbf{Z}(0), \mathbf{X}(0)) = (\mathbf{z}, \mathbf{x})]\right). \end{aligned} \tag{7.3.19}$$

Recall that $\theta(t) = \langle \boldsymbol{\theta}, \mathbf{X}(t)\rangle$. We further assume that $\pi(t) = \bar{\pi}(\mathbf{Z}(t), \mathbf{X}(t))$, the control processes (θ, π) are Markovian and feedback control processes.

Let $\mathcal{H}$ denote the space of functions $h(\cdot,\cdot) : \mathcal{T} \times \mathcal{R}^2 \times \mathcal{E} \to \mathbb{R}$ such that for each $\mathbf{x} \in \mathcal{E}$, $h(\cdot, \mathbf{x})$ is $\mathcal{C}^{1,2}(\mathcal{T} \times \mathcal{H}^2)$. Write $\mathbf{H}(\mathbf{z}) = (h(\mathbf{z}, \mathbf{e}_1), h(\mathbf{z}, \mathbf{e}_2), \ldots, h(\mathbf{z}, \mathbf{e}_N))' \in \mathbb{R}^N$.

Then, for each $(\theta, \pi) \in \Theta \times \mathcal{A}$, we define a Markovian regime-switching generator $\mathcal{L}^{\theta,\pi}$ acting on a function $h(\mathbf{z}, \mathbf{x}) \in \mathcal{H}$ for the Markovian process $\mathbf{Z}$ as below.

$$
\begin{aligned}
\mathcal{L}^{\theta,\pi}[h(\mathbf{z},\mathbf{x})] = & \frac{\partial h}{\partial u} + z_2[r(u) + (\mu(u) - r(u))\pi(\mathbf{z},\mathbf{x})]\frac{\partial h}{\partial z_2} \\
& + \theta^2(\mathbf{x})z_1^2\frac{\partial^2 h}{\partial z_1^2} + \frac{1}{2}z_2^2\pi^2(\mathbf{z},\mathbf{x})\sigma^2(u)\frac{\partial^2 h}{z_2^2} \\
& - \theta(\mathbf{x})\pi(\mathbf{z},\mathbf{x})z_1 z_2\sigma(u)\frac{\partial^2 h}{\partial z_1 \partial z_2} + \langle \mathbf{H}(\mathbf{z}), \Pi\mathbf{x}\rangle .
\end{aligned}
\tag{7.3.20}
$$

Lemma 7.3.1 *Suppose, for each* $\mathbf{x} \in \mathcal{E}$, $h(\mathbf{z}, \mathbf{x}) \in \mathcal{C}^{1,2}(\mathcal{T} \times \mathbb{R}^2)$. *Let* $\tau < \infty$ *be a stopping time. Assume further that* $h(\mathbf{Z}(u), \mathbf{X}(u))$ *and* $\mathcal{L}^{\theta,\pi} = [h(\mathbf{Z}(u), \mathbf{X}(u))]$ *are bounded on* $u \in [0, \tau]$. *Then*

$$
\begin{aligned}
& \mathbf{E}[h(\mathbf{Z}(\tau), \mathbf{X}(\tau))|(\mathbf{Z}(0), \mathbf{X}(0)) = (\mathbf{z}, \mathbf{x})] \\
& = h(\mathbf{z}, \mathbf{x}) + \mathbf{E}\left[\int_0^\tau \mathcal{L}^{\theta,\pi}[h(\mathbf{Z}(u), \mathbf{X}(u))du]|(\mathbf{Z}(0), \mathbf{X}(0)) = (\mathbf{z}, \mathbf{x})\right].
\end{aligned}
\tag{7.3.21}
$$

Proof The result follows by applying Itô's differentiation rule to $h(\mathbf{Z}(u), \mathbf{X}(u))$, using (7.3.20), integrating over the interval $[0, \tau]$, and conditioning on $(\mathbf{Z}(0), \mathbf{X}(0)) = (\mathbf{z}, \mathbf{x})$ under $\mathcal{P}$.

The following proposition presents the solution to the stochastic differential game between the representative agent and the market.

Proposition 7.3.2 *Let* $\mathcal{O} = (0, T) \times (0, \infty) \times (0, \infty)$. *Write* $\bar{\mathcal{O}}$, $\mathcal{C}^2(\mathcal{O})$ *and* $\mathcal{C}(\bar{\mathcal{O}})$ *for the closure of* $\mathcal{O}$, *the space of twice continuously differentiable functions on* $\mathcal{O}$ *and the space of continuously differentiable functions on* $\bar{\mathcal{O}}$, *respectively. Suppose there is a function* h *such that for each* $\mathbf{x} \in \mathcal{E}$, $h(\cdot, \mathbf{x}) \in \mathcal{C}^2(\mathcal{O}) \cap \mathcal{C}(\bar{\mathcal{O}})$, *and a Markovian control* $(\hat{\theta}, \hat{\pi}) \in \Theta \times \mathcal{A}$, *such that*

1. $\mathcal{L}^{\theta,\hat{\pi}(\mathbf{z},\mathbf{x})}[h(\mathbf{z}, \mathbf{x})] \geq 0$, *for all* $\theta \in \Theta$ *and* $(\mathbf{z}, \mathbf{x}) \in \mathcal{O} \times \mathcal{E}$;
2. $\mathcal{L}^{\hat{\theta}(\mathbf{x}),\hat{\pi}}[h(\mathbf{z}, \mathbf{x})] \leq 0$, *for all* $\pi \in \mathcal{A}$ *and* $(\mathbf{z}, \mathbf{x}) \in \mathcal{O} \times \mathcal{E}$;
3. $\mathcal{L}^{\hat{\theta}(\mathbf{x}),\hat{\pi}(\mathbf{z},\mathbf{x})}[h(\mathbf{z}, \mathbf{x})] = 0$, *for all* $(\mathbf{z}, \mathbf{x}) \in \mathcal{O} \times \mathcal{E}$;
4. $\lim_{t \to T^-} h(\mathbf{Z}(t), \mathbf{X}(t)) = Z_1(T) \cup (Z_2(T))$;
5. *Let* $\mathcal{K}$ *denote the set of stopping times* $\tau \leq T$. *The family* $\{\mathrm{h}(\mathbf{Z}(\tau), \mathbf{X}(\tau))|\tau \in \mathcal{K}\}$ *is uniformly integrable. Write, for each* $(\mathbf{z}, \mathbf{x}) \in \mathcal{O} \times \mathcal{E}$ *and* $(\theta, \pi) \in \Theta \times \mathcal{A}$,

$$
J^{\theta,\pi}(\mathbf{z}, \mathbf{x}) = \mathbf{E}[\Lambda^\theta(T)U(V^\pi(T))|(\mathbf{Z}(0), \mathbf{X}(0)) = (\mathbf{z}, \mathbf{x})]. \tag{7.3.22}
$$

then

$$
\begin{aligned}
h(\mathbf{z},\mathbf{x}) &= \Phi(\mathbf{z},\mathbf{x}) \\
&= \inf_{\theta\in\Theta}\left(\sup_{\pi\in\mathcal{A}}\mathbf{E}[\Lambda^{\theta}(T)U(V^{\pi}(T))|(\mathbf{Z}(0),\mathbf{X}(0))=(\mathbf{z},\mathbf{x})]\right) \\
&= \sup_{\pi\in\mathcal{A}} = \left(\inf_{\theta\in\Theta}\mathbf{E}[\Lambda^{\theta}(T)U(V^{\pi}(T))|(\mathbf{Z}(0),\mathbf{X}(0))=(\mathbf{z},\mathbf{x})]\right) \qquad (7.3.23)\\
&= \sup_{\pi\in\mathcal{A}} J^{\hat{\theta},\hat{\pi}}(\mathbf{z},\mathbf{x}) = \inf_{\theta\in\Theta} J^{\hat{\theta},\hat{\pi}}(\mathbf{z},\mathbf{x}) \\
&= J^{\hat{\theta},\hat{\pi}}(\mathbf{z},\mathbf{x}).
\end{aligned}
$$

and $(\hat{\theta},\hat{\pi})$ *is an optimal Markovian control.*

Proof The results can be proved by adapting to the proof of Theorem 3.2 in [17] and using the Dynkin formula presented in Lemma 7.3.1. So, we do not repeat it here.

Now, we solve the stochastic differential game. We suppose that the function h has the following form:

$$
h(\mathbf{z},\mathbf{x}) = \frac{z_1 z_2^{1-\zeta}(g(u,\mathbf{x}))^{1-\zeta}}{1-\zeta}, \forall(\mathbf{z},\mathbf{x}) \in \mathcal{O}\times\mathcal{E}, \qquad (7.3.24)
$$

where $g(T,\mathbf{x}) = 1$ does not vanish and $g(u,\mathbf{x})$, for each $u \in \mathcal{T}$, $\mathbf{x} \in \mathcal{E}$.

So, to determine the value function $h(\mathbf{z},\mathbf{x})$, we need to determine $g(u,\mathbf{x})$.

Recall that Conditions 1–3 of Proposition 7.3.2 are given by:

$$
\begin{aligned}
&\inf_{\theta\in\Theta}\mathcal{L}^{\theta,\hat{\pi}}[h(\mathbf{z},\mathbf{x})] = \mathcal{L}^{\hat{\theta},\hat{\pi}}[h(\mathbf{z},\mathbf{x})] = 0, \\
&\sup_{\pi\in\mathcal{A}}\mathcal{L}^{\hat{\theta},\pi}[h(\mathbf{z},\mathbf{x})] = \mathcal{L}^{\hat{\theta},\hat{\pi}}[h(\mathbf{z},\mathbf{x})] = 0.
\end{aligned} \qquad (7.3.25)
$$

By differentiating $\mathcal{L}^{\hat{\theta},\pi}[h(\mathbf{z},\mathbf{x})]$ in (7.3.20) with respect to π and setting the derivative equal to zero, we get the following first-order condition for $\hat{\pi}$ that maximizes $\mathcal{L}^{\hat{\theta},\pi}[h(\mathbf{z},\mathbf{x})]$ over all π:

$$
\begin{aligned}
&z_1 z_2^{1-\zeta}[g(u,\mathbf{x})]^{1-\zeta}(\mu(u) - r(u) - \zeta\hat{\pi}(\mathbf{z},\mathbf{x})\sigma^2(u) - \hat{\theta}(\mathbf{x})\sigma(u)) = 0, \\
&\forall(\mathbf{z},\mathbf{x}) \in \mathcal{O}\times\mathcal{E}.
\end{aligned} \qquad (7.3.26)
$$

Similarly, the first-order condition for $\widehat{\theta}$ that minimizes $\mathcal{L}^{\theta,\hat{\pi}}[h(\mathbf{z},\mathbf{x})]$ over all θ is given by:

$$-\hat{\pi}(\mathbf{z},\mathbf{x})\sigma(u)z_1z_2^{1-\zeta}[g(u,\mathbf{x})]^{1-\zeta}=0, \quad \forall(\mathbf{z},\mathbf{x})\in\mathcal{O}\times\mathcal{E}. \tag{7.3.27}$$

This implies that

$$\hat{\pi}(\mathbf{z},\mathbf{x})=0, \quad \forall(\mathbf{z},\mathbf{x})\in\mathcal{O}\times\mathcal{E}. \tag{7.3.28}$$

This means that in equilibrium, the optimal strategy of the representative agent is to invest all of his/her wealth in the money market account B, for every state of the economy.

Substituting (7.3.28) into (7.3.26),

$$z_1z_2^{1-\zeta}[g(u,\mathbf{x})]^{1-\zeta}(\mu(u)-r(u)-\hat{\theta}(\mathbf{x})\sigma(u))=0, \quad \forall(\mathbf{z},\mathbf{x})\in\mathcal{O}\times\mathcal{E}. \tag{7.3.29}$$

By noticing that $z_1z_2^{1-\zeta}>0, g(u,\mathbf{x})\neq 0$, for all $(\mathbf{z},\mathbf{x})\in\mathcal{O}\times\mathcal{E}$. Equation (7.3.29) implies that

$$\hat{\theta}(u)=\frac{\mu(u)-r(u)}{\sigma(u)}=\sum_{i=1}^{N}\left(\frac{\mu_i-r_i}{\sigma_i}\right)\langle\mathbf{X}(u),\mathbf{e}_i\rangle. \tag{7.3.30}$$

This implies that in equilibrium, the optimal strategy of the market is to choose the probability measure $\mathcal{P}^{\hat{\theta}}$ where $\hat{\theta}$ is given by (7.3.30).

$$\hat{\theta}(u)=\frac{\mu(u)-r(u)}{\sigma(u)}=\sum_{i=1}^{N}\left(\frac{\mu_i-r_i}{\sigma_i}\right)\langle\mathbf{X}(u),\mathbf{e}_i\rangle. \tag{7.3.30}$$

Now, Condition 3 in Proposition 7.3.2 states that

$$\mathcal{L}^{\hat{\theta},\hat{\pi}}[h(\mathbf{z},\mathbf{x})]=0. \tag{7.3.31}$$

Let $g_i(u)=g(u,\mathbf{e}_i)$, for each $i=1,2,\ldots,N$,$\mathbf{G}(u)=(g_1^{1-\zeta}(u),g_2^{1-\zeta}(u),\ldots,g_N^{1-\zeta}(u))'\in\mathbb{R}^N$. Then, from (7.3.20), (7.3.28), (7.3.30) and (7.3.31),

$$z_1z_2^{1-\zeta}[g(u,\mathbf{x})]^{-\zeta}\frac{dg(u,\mathbf{x})}{du}+z_1z_2^{1-\zeta}r(u)[g(u,\mathbf{x})]^{1-\zeta}+\frac{z_1z_2^{1-\zeta}}{1-\zeta}\langle\mathbf{G}(u),\Pi\mathbf{x}\rangle=0. \tag{7.3.32}$$

This implies that $g(u,\mathbf{x})$ satisfies the following Markovian regimes witching first-order backward ordinary differential equation (O.D.E.):

$$\frac{dg(u,\mathbf{x})}{du} + r(u)g(u,\mathbf{x}) + \frac{[g(u,\mathbf{x})]^{\zeta}}{1-\zeta}\langle \mathbf{G}(u), \Pi\mathbf{x}\rangle = 0, \tag{7.3.33}$$

with the terminal condition $g(T,\mathbf{x}) = 1$, for each $i = 1,2,\ldots,N$.

Equivalently, $g_i(u)$, $(i = 1,2,\ldots,N)$, satisfy the following system of coupled backward O.D.E.s:

$$\frac{dg_i(u)}{du} + r(u)g_i(u) + \frac{[g_i(u)]^{\zeta}}{1-\zeta}\langle \mathbf{G}(u), \Pi\mathbf{x}\rangle = 0, \tag{7.3.34}$$

with the terminal condition $g(T,\mathbf{x}) = 1$, for each $i = 1,2,\ldots,N$.

For each $i = 1,2,\ldots,N$, suppose $\pi_{ii} = -\lambda_i < 0$, where $\lambda_i > 0$. Note that $\sum_{j=1}^{N}\pi_{ij} = 0$. Then, the above system of O.D.E.s can be written as:

$$\frac{dg_i(u)}{du} + (r - \frac{\lambda_i}{1-\zeta})g_i(u) + \frac{[g_i(u)]^{\zeta}}{1-\zeta}\sum_{j=1,j\neq i}^{N}\pi_{ij}[g_j(u)]^{1-\zeta} = 0, \quad i = 1,2,\ldots,N. \tag{7.3.35}$$

If $r_i = r$, for all $i = 1,2,\ldots,N$, we can obtain closed-form solutions to the system of O.D.E.s (7.3.35). When $r_i = r$, for all $i = 1,2,\ldots,N$, the system of O.D. E.s becomes:

$$\frac{dg_i(u)}{du} + (r - \frac{\lambda_i}{1-\zeta})g_i(u) + \frac{[g_i(u)]^{\zeta}}{1-\zeta}\sum_{j=1,j\neq i}^{N}\pi_{ij}[g_j(u)]^{1-\zeta} = 0, \quad i = 1,2,\ldots,N. \tag{7.3.36}$$

We assume that the functions $g_i(u), i = 1,2,\ldots,N$, have the following form:

$$g_i(u) = e^{r(T-u)}. \tag{7.3.37}$$

So, $g_i(u)$ does not vanish, for each $u \in \mathcal{T}$, and $g_i(T) = 1$. It is not difficult to check that $g_i(u), i = 1,2,\ldots,N$, in (7.3.37) satisfy the system of O.D.E.s (7.3.36). Therefore, we obtain a closed-form expression for the value function $h(\mathbf{z},\mathbf{x})$ as follows:

$$h(\mathbf{z},\mathbf{x}) = \frac{z_1 z_2^{1-\zeta} e^{r(1-\zeta)(T-u)}}{1-\zeta}. \tag{7.3.38}$$

This form of the value function is consistent with the ones in Mataramvura and Øksendal (2007) [17] and in Øksendal and Sulem (2007) [26].

From (7.3.7) and (7.3.30), the density process corresponding to $\mathcal{P}^{\hat{\theta}}$ is:

$$\Lambda^{\hat{\theta}}(t) = \exp\left(\sum_{i=1}^{N}\left(\frac{r_i-\mu_i}{\sigma_i}\right)\int_0^t \langle \mathbf{X}(u), \mathbf{e}_i\rangle dw(u) - \frac{1}{2}\sum_{i=1}^{N}\left(\frac{r_i-\mu_i}{\sigma_i}\right)^2\int_0^t \langle \mathbf{X}(u), \mathbf{e}_i\rangle du\right). \tag{7.3.39}$$

Harrison and Kreps (1979) [27] and Harrison and Pliska (1981, 1983) [28, 29] developed an elegant mathematical theory for option valuation using the concept of a martingale. They established the relationship between the concept of a martingale and the absence of arbitrage. This is known as the fundamental theorem of asset pricing. It states that the absence of arbitrage is equivalent to the existence of an equivalent martingale measure under which discounted asset price processes are martingales. We call the existence of an equivalent martingale measure a martingale condition. The fundamental theorem of asset pricing was then extended by several authors, including Dybvig and Ross (1987) [30], Back and Pliska (1991) [31], Schachermayer (1992) [32] and Delbaen and Schachermayer (1994) [33], and others. Delbaen and Schachermayer (1994) [33] noted that the equivalence between the absence of arbitrage and the existence of an equivalent martingale measure is not always true in a continuous-time setting. They stressed that the term "essentially equivalent" instead of "equivalent" should be used to describe a martingale measure. Here, due to the presence of the additional source of uncertainty generated by the Markov chain $\mathbf{X}$, the martingale condition is given by considering an enlarged filtration.

Let $\tilde{\mathcal{G}}(t) = \mathcal{F}^{\mathbf{X}}(t) \vee \mathcal{F}^{Y}(t)$, for each $t \in \mathcal{T}$. Write $\tilde{\mathcal{G}} = \{\tilde{\mathcal{G}}(t) | t \in \mathcal{T}\}$. Let $\tilde{S}(t) = \exp(-\int_0^t r(u)du)S(t)$, which represents the discounted stock price at time $t \in \mathcal{T}$. Then, the martingale condition here is defined with respect to the enlarged filtration $\tilde{\mathcal{G}}$ and states that there is a probability measure $\mathcal{Q}$ equivalent to $\mathcal{P}$ such that,

$$\tilde{S}(u) = \mathrm{E}^{\mathcal{Q}}[\tilde{S}(t)|\tilde{\mathcal{G}}(u)], \quad \mathcal{P}\text{–a.s.}, \forall u, t \in \mathcal{T}, u \leq t,$$

where $\mathrm{E}^{\mathcal{Q}}[\cdot]$ represents expectation under $\mathcal{Q}$.

In particular, by letting $u = 0$, we require that for each $t \in \mathcal{T}$, the random variable

$$\mathbf{E}^{\mathcal{Q}}[\exp(-\int_0^t r(u)du)S(t)|\mathcal{F}^{\mathbf{X}}(t)] = S(0), \quad \mathcal{P}\text{–a.s.} \tag{7.3.40}$$

This means that, for each $t \in \mathcal{T}$, $R(t) := \mathbf{E}^{\mathcal{Q}}[\exp(-\int_0^t r(u)du)S(t)|\mathcal{F}^{\mathbf{X}}(t)]$ is an "almost surely" constant random variable under $\mathcal{P}$ such that $\mathcal{P}(R(t) = S(0)) = 1$, for each $t \in \mathcal{T}$. The condition is presented in Elliott et al. (2005) [34].

From (7.3.11), (7.3.12) and (7.3.30), the evolution of the stock price process under $\mathcal{P}^{\hat{\theta}}$ is given by:

$$dS(t) = r(t)S(t)dt + \sigma(t)S(t)dw^{\hat{\theta}}(t). \tag{7.3.41}$$

Hence

$$\mathbf{E}^{\hat{\theta}}\left[\exp\left(-\int_0^t r(u)du\right)S(t)\middle|\mathcal{F}^{\mathbf{X}}(t)\right] = S(0), \quad \mathcal{P}\text{–a.s.} \tag{7.3.42}$$

In other words, the condition (7.3.29) is satisfied under the probability measure $\mathcal{P}^{\hat{\theta}}$ deduced from an equilibrium state of the stochastic differential game between the representative agent and the market.

The Esscher transform has a long and remarkable history in actuarial science. It was first introduced to the actuarial science literature by Esscher (1932) [35]. It has different applications in actuarial science, such as premium calculation and approximation to the aggregate claims distribution. In the two landmark papers by Bühlmann (1980, 1984) [36, 37], he established an economic equilibrium premium principle in the sense of the Pareto optimal risk exchange under a pure exchange economy and a link between the economic premium principle and the premium rule determined by the Esscher transform. Gerber and Shiu (1994) [38] pioneered the use of the Esscher transform for option pricing. Their work opened up many research opportunities to further explore the interplay between actuarial and financial pricing. Elliott et al. (2005) [39] considered a regime-switching version of the Esscher transform to determine an equivalent martingale measure in the context of a Markovian regime-switching Black–Scholes–Merton market. In the sequel, we demonstrate that an equivalent martingale measure chosen by the regime-switching Esscher transform considered in Elliott et al. (2005) is identical to $\mathcal{P}^{\hat{\theta}}$.

Define, for each $t \in \mathcal{T}$, the regime-switching Esscher parameter $\beta(t)$ at time t as below:

$$\beta(t) = \langle \boldsymbol{\beta}, \mathbf{X}(t) \rangle, \tag{7.3.42}$$

where, $\boldsymbol{\beta} := (\beta_1, \beta_2, \ldots, \beta_N) \in \mathbb{R}^N$, $\beta_{(N)} = \max_{1 \leq i \leq N} \beta_i < \infty$.

Write $(\beta \cdot w)(t) = \int_0^t \beta(u)dw(u)$, for each $t \in \mathcal{T}$. Define a process $\Lambda^\beta = \{\Lambda^\beta | t \in \mathcal{T}\}$ on $(\Omega, \mathcal{F}, \mathcal{P})$ as below:

$$\Lambda^\beta(t) = \frac{e^{-(\beta\cdot w)(t)}}{\mathbf{E}[e^{-(\beta\cdot w)(t)}|\mathcal{F}^{\mathbf{X}}(t)]}. \tag{7.3.43}$$

By Itô's differentiation rule,

$$e^{-(\beta\cdot w)(t)} = 1 - \int_0^t e^{-(\beta\cdot w)(t)}\beta(u)dw(u) + \frac{1}{2}\int_0^t e^{-(\beta\cdot w)(u)}\beta^2(u)du. \tag{7.3.44}$$

Conditioning on $\mathcal{F}^{\mathbf{X}}(t)$ under $\mathcal{P}$,

$$\mathbf{E}[e^{-(\beta\cdot w)(t)}|\mathcal{F}^{\mathbf{X}}(t)] = 1 + \frac{1}{2}\int_0^t \mathbf{E}[\mathrm{e}^{-(\beta\cdot w)(t)}|\mathcal{F}^{\mathbf{X}}(t)]\beta^2(u)du. \tag{7.3.45}$$

Solving (7.3.45),

$$\mathbf{E}[\mathrm{e}^{-(\beta\cdot w)(t)}|\mathcal{F}^{\mathbf{X}}(t)] = \exp\left(\frac{1}{2}\int_0^t \beta^2(u)du\right). \tag{7.3.46}$$

Then

$$\Lambda^\beta(t) = \exp\left(-\int_0^t \beta(u)dw(u) - \frac{1}{2}\int_0^t \beta^2(u)du\right). \tag{7.3.47}$$

So

$$d\Lambda^\beta(t) = -\beta(t)\Lambda^\beta(t)dw(u). \tag{7.3.48}$$

Here, Λ^β is a $(\mathcal{G},\mathcal{P})$-local-martingale.

Note that the Novikov condition $\mathbf{E}\left[\exp\left(-\frac{1}{2}\int_0^T \beta^2(u)du\right)\right] < \infty$ is satisfied. Then, Λ^β is a $(\mathcal{G},\mathcal{P})$-martingale, and, hence

$$\mathbf{E}[\Lambda^\beta(t)] = 1, \quad t \in \mathcal{T}. \tag{7.3.49}$$

Similar to Elliott et al. (2005), the regime-switching Esscher transform $\mathcal{P}^\beta \sim \mathcal{P}$ on $\mathcal{G}(T)$ is defined as:

$$\frac{d\mathcal{P}^\beta}{d\mathcal{P}} = \Lambda^\beta(T). \tag{7.3.50}$$

We then seek $\beta = \hat{\beta}$, the risk-neutral Esscher process, such that

$$\mathbf{E}^{\hat{\beta}}\left[\exp\left(-\int_0^t r(u)du\right)S(t)|\mathcal{F}^{\mathbf{X}}(t)\right] = S(0), \quad \mathcal{P}\text{−a.s.} \tag{7.3.51}$$

By the Bayes' rule,

$$\begin{aligned}
&\mathbf{E}^{\hat{\beta}}\left[\exp\left(-\int_0^t r(u)du\right)S(t)|\mathcal{F}^{\mathbf{X}}(t)\right]\\
&\quad=\mathbf{E}\left[\Lambda^{\hat{\beta}}(t)\exp\left(-\int_0^t r(u)du\right)S(t)|\mathcal{F}^{\mathbf{X}}(t)\right]\\
&\quad=S(0)\exp\left(\int_0^t(\mu(u)-r(u)-\hat{\beta}(u)\sigma(u))du\right).
\end{aligned}\tag{7.3.52}$$

Then,

$$\int_0^t(\mu(u)-r(u)-\hat{\beta}(u)\sigma(u))du=0,\quad\forall t\in\mathcal{T}\tag{7.3.53}$$

This implies that

$$\hat{\beta}(u)=\frac{\mu(u)-r(u)}{\sigma(u)}=\sum_{i=1}^N\left(\frac{\mu_i-r_i}{\sigma_i}\right)\langle\mathbf{X}(u),\mathbf{e}_i\rangle=\hat{\theta}(u).\tag{7.3.54}$$

Here, the process $\hat{\theta}$ characterizing the probability measure $\mathcal{P}^{\hat{\theta}}$ from the equilibrium state of the stochastic differential game and the risk-neutral Esscher process $\hat{\beta}$ are identical.

Then, from (7.3.47) and (7.3.54),

$$\begin{aligned}
\Lambda^{\hat{\beta}}(t)&=\exp\Big(\sum_{i=1}^N\left(\frac{r_i-\mu_i}{\sigma_i}\right)\int_0^t\langle\mathbf{X}(u),\mathbf{e}_i\rangle dw(u)\\
&\qquad-\frac{1}{2}\sum_{i=1}^N\left(\frac{r_i-\mu_i}{\sigma_i}\right)^2\int_0^t\langle\mathbf{X}(u),\mathbf{e}_i\rangle du\Big)\\
&=\Lambda^{\hat{\theta}}(t).
\end{aligned}\tag{7.3.55}$$

Hence, $\mathcal{P}^{\hat{\beta}}$ is identical to $\mathcal{P}^{\hat{\theta}}$.

By Girsanov's theorem,

$$w^{\hat{\beta}}(t)=w(t)-\int_0^t\left(\frac{r(u)-\mu(u)}{\sigma(u)}\right)du\tag{7.3.56}$$

is a standard Brownian motion with respect to $\mathcal{G}$ under $\mathcal{P}^{\hat{\beta}}$.

Under $\mathcal{P}^{\hat{\beta}}$ the evolution of the stock price process is governed by:

$$dS(t)=r(t)S(t)dt+\sigma(t)S(t)dw^{\hat{\beta}}(t).\tag{7.3.57}$$

This coincides with the stock price dynamics under $\mathcal{P}^{\hat{\theta}}$.

7.4 The Optimal Investment Game of the Insurance Company

7.4.1 The Market Model

We consider a continuous-time, Markov, regime-switching, economic model with a bond and a capital market index, or a share index. The following assumptions are then imposed:

(1) the bond and the index can be traded continuously over time;
(2) there is no transaction cost or tax involved in trading;
(3) the bond and the index are liquid;
(4) any fractional units of the bond and the index can be traded.

Suppose $(\Omega, \mathcal{F}, \mathcal{P})$ is a complete probability space, where P represents a reference probability measure from which a family of real-world probability measures are generated. The measure $\mathcal{P}$ is the probability measure characterizing a reference model. We assume that $(\Omega, \mathcal{F}, \mathcal{P})$ is rich enough to describe uncertainties. All economic activities take place in a time horizon denoted by $\mathcal{T}$, where $\mathcal{T} := [0, T]$, for $T < \infty$. We model the evolution of the state of an economy over time by a continuous-time, finite-state, observable Markov chain $\mathbf{X} := \{\mathbf{X}(t) | t \in \mathcal{T}\}$ on $(\Omega, \mathcal{F}, \mathcal{P})$ taking values in the state space $\mathcal{S} := \{s_1, s_2, \ldots, s_N\}$, where $N \geq 2$. Without loss of generality, we adopt the formalism introduced by Elliott et al. (1994) and identify the state space of the chain by a set of unit basis vectors $\mathcal{E} := \{\mathbf{e}_1, \mathbf{e}_2, \ldots, \mathbf{e}_N\} \in \mathbb{R}^N$. Here the jth component of $\mathbf{e}_i$ is the Kronecker delta, denoted as δ_{ij}, for each $i, j = 1, 2, \ldots, N$. We call the set $\mathcal{E}$ the canonical state space of the chain $\mathbf{X}$.

Let $\mathbf{A} = [a_{ij}]_{i,j=1,2,\ldots,N}$ denote the generator of the chain $\mathbf{X}$ under the real-world probability measure $\mathcal{P}$. For each $i, j = 1, 2, \ldots, N$, a_{ji} is the constant, instantaneous, intensity of the transition of the chain $\mathbf{X}$from state i P to state j. Note that $a_{ji} \geq 0$, for $i \neq j$, and that $\sum_{j=1}^{N} a_{ji} = 0$, so $a_{ii} \leq 0$. Here for each $i, j = 1, 2, \ldots, N$, with $i \neq j$, we suppose that a_{ji}40, so a_{ji}50. We assume that the chain $\mathbf{X}$ is irreducible. Then there is a unique stationarydistribution for the chain. Let $\mathbf{y}'$ denote the transpose of a matrix, or a vector, $\mathbf{y}$. With the canonical state space of the chain, Elliott et al. [19] (1994) derived the following semi-martingaledynamics for $\mathbf{X}$:

$$\mathbf{X}(t) = \mathbf{X}(0) + \int_0^t \mathbf{A}'\mathbf{X}(u) du + \mathbf{M}(t),$$

Here $\{\mathbf{M}(t) | t \in \mathcal{T}\}$ is an $\mathbb{R}^N$-valued martingale with respect to the right-continuous, $\mathcal{P}$-completed, filtration generated by $\mathbf{X}$ under the measure $\mathcal{P}$.

Now we specify the price processes of the bond and the share index. Let $r(t)$ be the interest rate of the bond at time t. Then the chain determines $r(t)$ as

$$r(t) = \langle \mathbf{r}, \mathbf{X}(t) \rangle,$$

Here $\mathbf{r} := (r_1, r_2, \ldots, r_N)' \in \mathbb{R}^N$ with $r_i > 0$, for each $i = 1, 2, \ldots, N$. Then the price process $B := \{B(t) | t \in \mathcal{T}\}$ of the bond evolves over time as

$$B(t) = \exp\left(\int_0^t r(\mathrm{u}) du \right), \quad t \in \mathcal{T}, \quad B(0) = 1.$$

Then the chain determines the appreciation rate $\mu(t)$ and the volatility $\sigma(t)$ as

$$\mu(t) = \langle \boldsymbol{\mu}, \mathbf{X}(t) \rangle,$$
$$\sigma(t) = \langle \boldsymbol{\sigma}, \mathbf{X}(t) \rangle.$$

Here $\boldsymbol{\mu} := (\mu_1, \mu_2, \ldots, \mu_N)' \in \mathbb{R}^N$, $\boldsymbol{\sigma} := (\sigma_1, \sigma_2, \ldots, \sigma_N)' \in \mathbb{R}^N$ with $\mu_i > r_i$, $\sigma_i > 0$ for each $i = 1, 2, \ldots, N$.

Suppose $w := \{w(t) | t \in \mathcal{T}\}$ denotes a standard Brownian motion on $(\Omega, \mathcal{F}, \mathcal{P})$ with respect to its rightcontinuous, $\mathcal{P}$-completed, filtration $\mathcal{F}^w := \{\mathcal{F}^w(t) | t \in \mathcal{T}\}$.

Then the evolution of the price process of the index is governed by the following Markov, regime-switching, Geometric Brownian Motion (GBM):

$$dS(t) = \mu(t) S(t) dt + \sigma(t) S(t) dw(t), \; S(0) = s_0.$$

In the sequel, we present two specifications for the insurance risk process. The first specification is a Markov, regime-switching, compound Poisson process, while the second specification is a Markov, regime-switching, diffusion process. In each case, we derive the surplus process of an insurance company.

7.4.1.1 A Markov Regime-Switching Random Measure for Insurance Claims

Here we present a Markov, regime-switching, random measure for the aggregate insurance claims process.

Let $Z := \{Z(t) | t \in \mathcal{T}\}$ denote a real-valued, Markov, regime-switching, pure jump process on $(\Omega, \mathcal{F}, \mathcal{P})$. Here $Z(t)$ is the aggregate amount of claims up to and including time t. Suppose, for each $\mathrm{u} \in \mathcal{T}$, $\Delta Z(\mathrm{u}) := Z(\mathrm{u}) - Z(\mathrm{u}-)$, the jump size of the process Z at time u. Then

$$Z(t) = \sum_{0<\mathrm{u}<t} \Delta Z(\mathrm{u}), \quad Z(0) = 0, \quad \mathcal{P}\text{-a.s.}, \; \mathrm{t} \in \mathcal{T}. \tag{7.4.1}$$

Suppose the state space of claim size, denoted as $\mathcal{Z} \subset (0, \infty)$. Let $\mathcal{M}$ denote the product space $\mathcal{T} \times \mathcal{Z}$ of claim arrival time and claim size. Define a random measure $\gamma(\cdot, \cdot)$ on the product space $\mathcal{M}$, which selects claim arrivals and sizes

$z := Z(\mathrm{u}) - Z(\mathrm{u}-)$, Indeed, the random measure can be written as a sum of random delta functions, that is

$$\gamma(dz, d\mathrm{u}) = \sum_{k \geq 1} \delta_{(\Delta Z(T_k), T_k)}(dz, d\mathrm{u}) I_{\{\Delta Z(T_k) \neq 0, T_k < \infty\}}. \tag{7.4.2}$$

Here T_k is the arrival time of the kth claim, $\Delta Z(T_k)$ is the amount of the kth claim at the time epoch T_k, $\delta_{(\Delta Z(T_k), T_k)}(\cdot, \cdot)$ is the random delta functions at the point $(\Delta Z(T_k), T_k) \in \mathcal{Z} \times \mathcal{T}$, and I_E is the indicator function of an event E

For suitable integrands $f : (\Omega \times \mathcal{Z} \times \mathcal{T}) \to \Re$

$$\int_0^t \int_0^\infty f(\omega, z, \mathrm{u}) \gamma(dz, d\mathrm{u}) = \sum_{T_k \leq t} f(\omega, \Delta Z(T_k), T_k), \quad t \in \mathcal{T}. \tag{7.4.3}$$

The aggregate insurance claims process Z can then be written as.

$$Z(t) = \int_0^t \int_0^\infty z\gamma(dz, d\mathrm{u}), \quad t \in \mathcal{T}. \tag{7.4.4}$$

Define, for each $t \in \mathcal{T}$,

$$N(t) = \int_0^t \int_0^\infty \gamma(dz, d\mathrm{u}). \tag{7.4.5}$$

So $N(t)$ counts the number of claim arrivals up to time t.

We assume that under $\mathcal{P}$, $N := \{N(t) | t \in \mathcal{T}\}$ is conditional Poisson process on $(\Omega, \mathcal{F}, \mathcal{P})$ with intensity modulated by the chain $\mathbf{X}$ as

$$\lambda(t) := \langle \lambda, \mathbf{X}(t) \rangle, \quad t \in \mathcal{T}. \tag{7.4.6}$$

Here $\lambda := (\lambda_1, \lambda_2, \ldots, \lambda_N)' \in \mathbb{R}^N$ with $\lambda_i > 0$, and λ_i is the jump intensity of N when the economy is in the ith state, for each $i = 1, 2, \ldots, N$.

Now we specify the distribution of claim sizes. For each $i = 1, 2, \ldots, N$, let $F_i(z)$ denote a probability distribution function of the claim size $z := Z(\mathrm{u}) - Z(\mathrm{u}-)$ when $\mathbf{X}(\mathrm{u}-) = \mathbf{e}_i$. Then, the compensator of the Markov, regime-switching, random measure $\gamma(\cdot, \cdot)$ under $\mathcal{P}$ is

$$\nu_{\mathbf{X}(u-)}(dz, du) := \sum_{i=1}^N \langle \mathbf{X}(\mathrm{u}-), \mathbf{e}_i \rangle \lambda_i F_i(dz) d\mathrm{u} \tag{7.4.7}$$

so a compensated version of the Markov, regime-switching, random measure, denoted by $\bar{\gamma}(\cdot, \cdot)$, is

$$\overline{\gamma}(dz, du) := \gamma(dz, du) - \nu_{\mathbf{X}(u-1)}(dz, du). \tag{7.4.8}$$

Let $p(t)$ denote the premium rate at time t, for each $t \in \mathcal{T}$. Then we suppose that the chain $\mathbf{X}$ determines $p(t)$ as

$$p(t) := \langle \mathbf{p}, \mathbf{X}(t) \rangle, \quad t \in \mathcal{T}. \tag{7.4.9}$$

Here $\mathbf{p} := (p_1, p_2, \ldots, p_N) \in \mathbb{R}^N$, $p_i > 0$.

Suppose $R := \{R(t) | t \in \mathcal{T}\}$ denotes the surplus process of the insurance company without investment. Then

$$\begin{aligned} R(t) &:= u + \int_0^t p(u)du - Z(t) \\ &= u + \sum_{i=1}^{N} p_i \mathcal{J}_i(t) - \int_0^t \int_0^\infty z\gamma(dz, du), \quad t \in \mathcal{T}, \end{aligned} \tag{7.4.10}$$

where the initial surplus $R(0) = u$; for each $i = 1, 2, \ldots, N$ and each $t \in \mathcal{T}$, $\mathcal{J}_i(t)$ is the occupation time of the chain $\mathbf{X}$ in state $\mathbf{e}_i$ up to time t, that is

$$\mathcal{J}_i(t) := \int_0^t \langle \mathbf{X}(u), \mathbf{e}_i \rangle du.$$

We now derive the surplus process of an insurance company which invests its surplus in the bond and the hare index. Firstly, we specify the information structure o the model Let $\mathcal{F}^Z := \{\mathcal{F}^Z(t) | t \in \mathcal{T}\}$ denote the right continuous, $\mathcal{P}$-completed, filtration generated by the history of the insurance claims process Z, that is $\mathcal{F}^Z(t)$ is the P-augmentation of the σ-field generated by the insurance claims process Z up to and including time t. Define, for each $t \in \mathcal{T}$, the enlarged σ-algebra $\mathcal{G}(t) = \mathcal{F}^w(t) \vee \mathcal{F}^Z(t) \vee \mathcal{F}^{\mathbf{X}}(t)$, the minimal σ-field generated by $\mathcal{F}^w(t)$, $\mathcal{F}^Z(t)$, $\mathcal{F}^{\mathbf{X}}(t)$. Write $\mathcal{G} := \{\mathcal{G}(t) | t \in \mathcal{T}\}$. We assume that the insurance company can observe $\mathcal{G}(t)$ at each time $t \in \mathcal{T}$. Consequently, the company observes the values of the share index, insurance claims and economic conditions.

Suppose the insurance company invests the amount of $\pi(t)$ in the share index at time t, for each $t \in \mathcal{T}$, we write $V(t) = V^\pi(t)$ unless otherwise stated. Since the money used for investment in the share index comes solely from the surplus process, we must have $\pi(t) < V(t)$, for each $t \in \mathcal{T}$. Consequently, for each $t \in \mathcal{T}$, the amount invested in the bond at time t is $V(t) - \pi(t)$. Then, the surplus process of the insurance company with investment evolves over time as

$$\begin{aligned} dV(t) &= [p(t) + r(t)V(t) + \pi(t)(\mu(t) - r(t))]dt \\ &\quad + \sigma(t)\pi(t)dw(t) - \int_0^\infty z\gamma(dz, dt), \\ V(0) &= v. \end{aligned} \tag{7.4.11}$$

Here we say that a portfolio process π is admissible if it satisfies the following conditions:

(1) π is $\mathcal{G}$-progressively measurable;
(2) the stochastic differential equation for the surplus dynamics has a unique strong solution;:

(3) $$\sum_{i=1}^{N} \int_0^T \left\{ |p_i + r_i V_i + \pi(t)(\mu_i - r_i)| + \sigma_i^2 \pi^2(t) + \lambda_i \int_0^\infty zF_i(dz) \right\} dt < \infty,$$
$\mathcal{P}$–a.s.;

(4) $V(t) \geq 0$ for all $t \in \mathcal{T}$, $\mathcal{P}$–a.s.

The first condition states that the insurance company decides the amount invested in the share index based on the current and past price information, observations about insurance risk process, and economic information. The second and third conditions are technical conditions. The last condition is the solvency condition of the insurance company. We denote here the set of all admissible portfolio processes of the insurance company by $\mathcal{A}$.

7.4.1.2 A Markov Regime-Switching Diffusion Process for Insurance Risk

Now we introduce a Markov, regime-switching, diffusion risk process. Let $\sigma_z(t)$ denote the instantaneous volatility of the aggregate insurance claims process Z at time t, for each $t \in \mathcal{T}$.

The chain X determines $\sigma_z(t)$ as

$$\sigma_z(t) = \langle \boldsymbol{\sigma}_z, \mathbf{X}(t) \rangle.$$

Here $\boldsymbol{\sigma}_z = (\sigma_{z1}, \sigma_{z2}, \ldots, \sigma_{zN})' \in \mathbb{R}^N$ with $\sigma_{zi} > 0$; for each $i = 1, 2, \ldots, N$. σ_{zi} represents the uncertainty of the surplus process of the insurance company without investment when the economy is in the ith state. The model proposed here can accommodate the situation when the insurance company faces different levels of uncertainty in different economic conditions. For example, when the economy is in a ‘Bad’ state, the insurance company faces a higher level of uncertainty in its surplus process than when the economy is in a ‘Good’ state.

In practice, insurance risk processes and financial price processes may be correlated. Their correlation may depend on the economic condition. The correlation between insurance risk processes and financial price processes in a ‘Bad’ economy may be higher than that in a ‘Good’ economy. Let $\rho(t)$ denote the instantaneous correlation coefficient between the random shock of the price process of the index and that of the insurance risk process at time t, for each $t \in \mathcal{T}$. Then we assume that the chain $\mathbf{X}$ determines $\rho(t)$ as

$$\rho(t) := \langle \boldsymbol{\rho}, \mathbf{X}(t) \rangle, \quad t \in \mathcal{T}.$$

Here $\boldsymbol{\rho} = (\rho_1, \rho_2, \ldots, \rho_N)' \in \mathbb{R}^N$ with $\rho_i \in (-1, 1)$ for each $i = 1, 2, \ldots, N$, when the economy is in state i, ρ_i is the correlation coefficient between the random shock of the insurance risk process and that of the price process of the index.

Let $w^z = \{w^z(t) | t \in \mathcal{T}\}$ denote a standard Brownian motion on $(\Omega, \mathcal{F}, \mathcal{P})$, which represents the 'noise' in the insurance risk process. We allow the flexibility that $\{w^z(t) | t \in \mathcal{T}\}$ and $\{w(t) | t \in \mathcal{T}\}$ are correlated with instantaneous correlation coefficient $\rho(t)$ at time t. Then the aggregate insurance claims process $\bar{Z} = \{\bar{Z}(t) | t \in \mathcal{T}\}$ follows

$$\bar{Z}(t) = \int_0^t \sigma^z(\mathrm{u}) dw^z(\mathrm{u}), \quad t \in \mathcal{T}. \tag{7.4.12}$$

Let $\bar{R} := \{\bar{R}(t) | t \in \mathcal{T}\}$ denote the surplus process of the insurance company without investment. Then

$$\bar{R}(t) = u + \int_0^t p(\mathrm{u}) d\mathrm{u} - \bar{Z}(t), \tag{7.4.13}$$

where, as before, it is given by (7.4.10).

Let $\bar{\mathcal{F}}^{\bar{z}} := \left\{ \bar{\mathcal{F}}^{\bar{z}}(t) | t \in \mathcal{T} \right\}$ denote the right-continuous, $\mathcal{P}$-completed, filtration generated by the insurance claims process $\bar{Z} = \{\bar{Z}(t) | t \in \mathcal{T}\}$. For each $t \geq 0$, define an enlarged σ-field $\bar{\mathcal{G}}(t) = \bar{\mathcal{F}}^{\bar{Z}}(t) \vee \mathcal{F}^w(t) \vee \mathcal{F}^{\mathbf{X}}(t)$, Write $\bar{\mathcal{G}} = \{\bar{\mathcal{G}}(t) | t \in \mathcal{T}\}$. Again, we assume that the insurance company can access to $\bar{\mathcal{G}}(t)$ at each time $t \in \mathcal{T}$.

For each $t \in \mathcal{T}$, let $\bar{\pi}(t)$ denote the amount the insurance company invests in the index at time t. Let $\{\bar{V}^{\bar{\pi}}(t) | t \in \mathcal{T}\}$ denote the surplus process of the company associated with a portfolio process π. Again to simplify the notation, we write $\bar{V}(t) = \bar{V}^{\bar{\pi}}(t)$, for each $t \in \mathcal{T}$. We must also have that $\pi(t) < V(t)$, for each $t \in \mathcal{T}$. The amount invested in the bond at time t is $V(t) - \pi(t)$, for each $t \in \mathcal{T}$. Then the surplus process of the insurance company with investment evolves over time as

$$d\bar{V}(t) = [p(t) + r(t)\bar{V}(t) + \bar{\pi}(t)(\mu(t) - r(t))]dt + \sigma(t)\bar{\pi}(t)dw(t) - \sigma_z(t)dw^z(t). \tag{7.4.14}$$

Similarly to the last subsection, we say that a portfolio process $\bar{\pi}(t)$ is admissible if it satisfies the following conditions:

(1) π is $\bar{\mathcal{G}}$-progressively measurable;

(2) the surplus process has a unique strong solution;
(3) $\int_0^T \bar{\pi}^2(t)dt < \infty$, $\mathcal{P}$−a.s.
(4) $\overline{V}(t) \geq 0$, for all $t \in \mathcal{T}$, $\mathcal{P}$−a.s.

We denote the space of admissible portfolio processes by $\bar{\mathcal{A}}$.

7.4.2 *Optimal Investment Problems*

In this section, we discuss two optimal investment problems. The first problem is to maximize the minimal expected exponential utility of terminal wealth over a family of real-world probability measures when the insurance risk process is governed by the Markov, regime-switching, random measure.

The second problem is to maximize the minimal survival probability over another family of real-world probability measures when the Markov, regime-switching, diffusion-based risk process is considered. We adopt here a robust approach to model risk, or uncertainty. We formulate the two investment problems as two zero-sum, two-player, stochastic differential games between the insurance company and the market. One of the games is a finite-horizon game and the other is an infinite-horizon one.

7.4.2.1 Maximizing the Minimal Expected Exponential Utility

Firstly, we introduce processes $\{\theta = \theta(t)|t \in \mathcal{T}\}$ which parameterize the family of real-world probability measures. Suppose the process θ satisfies the following conditions:

(1) θ is $\mathcal{G}$-progressively measurable;
(2) $\theta(t) := \theta(t, \omega) \leq 1$, for a. a.$(t, \omega) \in \mathcal{T} \times \Omega$;
(3) $\int_0^T \theta^2(t)dt < \infty$, $\mathcal{P}$−a.s.

We denote the space of all such processes by Θ.

Define, for each $\theta \in \Theta$, a real-valued, $\mathcal{G}$-adapted, process $\Lambda^\theta := \{\Lambda^\theta(t)|t \in \mathcal{T}\}$ on $(\Omega, \mathcal{F}, \mathcal{P})$ by

$$\Lambda^\theta(t) := \exp\bigg(-\int_0^t \theta(\mathrm{u})dw(\mathrm{u}) - \frac{1}{2}\int_0^t \theta^2(\mathrm{u})d\mathrm{u} + \int_0^t\int_0^\infty \ln(1-\theta(\mathrm{u}))\bar{\gamma}(dz, d\mathrm{u}) + \int_0^t\int_0^\infty \{\ln(1-\theta(\mathrm{u})) + \theta(\mathrm{u})\}\nu(dz, d\mathrm{u})\bigg). \tag{7.4.15}$$

Applying Itô's differentiation rule to Λ^θ gives

$$\begin{aligned} d\Lambda^{\theta}(t) &= \Lambda^{\theta}(t-)(-\theta(t)dw(t) - \int_0^{\infty} \theta(t)\bar{\gamma}(dz,dt)), \\ \Lambda^{\theta}(t) &= 1, \quad \mathcal{P}\text{-a.s.} \end{aligned} \tag{7.4.16}$$

So Λ^{θ} is a $(\mathcal{G},\mathcal{P})$-(local)-martingale. We suppose that $\theta \in \Theta$ is such that Λ^{θ} is a $(\mathcal{G},\mathcal{P})$-martingale. Consequently, $\mathrm{E}\left[\Lambda^{\theta}(T)\right] = 1$.

We now introduce a density process for the measure change of the regime-switching Markov chain. For each $t \in \mathcal{T}$, let $\mathcal{F}(t) := \mathcal{F}^{w}(t) \vee \mathcal{F}^{Z}(t)$, the minimal σ-field generated by $\mathcal{F}^{w}(t)$ and $\mathcal{F}^{Z}(t)$. Write $\mathcal{F} = \{\mathcal{F}(t)|t \in \mathcal{T}\}$. For each $i,j = 1,2,\ldots,N$, let $\{c_{ij}(t)|t \in \mathcal{T}\}$ be a real-valued, $\mathcal{F}$-predictable, bounded stochastic process $\{c_{ij}(t)|t \in \mathcal{T}\}$ on $(\Omega,\mathcal{F},\mathcal{P})$ such that, for each $t \in \mathcal{T}$,

(1) $c_{ij}(t) \geq 0$, for $i \neq j$; and
(2) $\sum_{i=1}^{N} c_{ij}(t) = 0$, so $c_{ii}(t) \leq 0$.

Then a second family of rate matrices, $\mathbf{C} = \{\mathbf{C}(t)|t \in \mathcal{T}\}$, can be defined by $\mathbf{C}(t) = [c_{ij}(t)]_{i=1,2,\ldots,N}$.

We wish to introduce a new probability measure under which $\mathbf{C}$ is a family of rate matrices of the chain $\mathbf{X}$ with indexed set $\mathcal{T}$. We adopt a version of Girsanov's transform for the Markov chain considered by Dufour and Elliott (1999) to define the new probability measure.

Firstly, for each $t \in \mathcal{T}$, we define the following matrix:

$$\mathbf{D}^{\mathbf{C}}(t) := [c_{ij}(t)/a_{ij}(t)]_{i,j=1,2,\ldots,N} = \left[d_{ij}^{\mathbf{C}}(t)\right].$$

Note that $a_{ij}(t) > 0$, for each $t \in \mathcal{T}$, so $\mathbf{D}(t)$ is well-defined.
For each $t \in \mathcal{T}$, let $\mathbf{d}^{\mathbf{C}}(t) := (d_{11}^{\mathbf{C}}(t), d_{22}^{\mathbf{C}}(t), \ldots, d_{NN}^{\mathbf{C}}(t))' \in \mathbb{R}^N$.
Write, for each $t \in \mathcal{T}$,

$$\mathbf{D}_0^{\mathbf{C}}(t) = \mathbf{D}^{\mathbf{C}}(t) - \mathbf{diag}(\mathbf{d}^{\mathbf{C}}(t)).$$

Here $\mathbf{diag}(\mathbf{y})$ is a diagonal matrix with diagonal elements given by the vector $\mathbf{y}$. Consequently, $\mathbf{D}_0^{\mathbf{C}}(t)$ is the matrix $\mathbf{D}^{\mathbf{C}}(t)$ with its diagonal elements being taken out.

Consider the vector-valued counting process, $\mathbf{N}^{\mathbf{X}} : \{\mathbf{N}^{\mathbf{X}}(t)|t \in \mathcal{T}\}$ on $(\Omega,\mathcal{F},\mathcal{P})$, where for each $t \in \mathcal{T}$, $\mathbf{N}^{\mathbf{X}}(t) = (N_1^{\mathbf{X}}(t), N_1^{\mathbf{X}}(t), \ldots, N_N^{\mathbf{X}}(t))' \in \mathbb{R}^N$ and $N_j^{\mathbf{X}}(t)$ counts the number of jumps of the chain $\mathbf{X}$ to state $\mathbf{e}_j$ up to time t, for each $j = 1,2,\ldots,N$. Then it is not difficult to check that $\mathbf{N}$ has the following semi-martingale dynamics:

$$\begin{aligned}\mathbf{N}^{\mathbf{X}}(t) &= \int_0^t (\mathbf{I} - \mathbf{diag}(\mathbf{X}(\mathrm{u}-)))' d\mathbf{X}(\mathrm{u}) \\ &= \mathbf{N}^{\mathbf{X}}(0) + \int_0^t (\mathbf{I} - \mathbf{diag}(\mathbf{X}(\mathrm{u}-)))' \mathbf{A}(t)\mathbf{X}(t)dt + \int_0^t (\mathbf{I} - \mathbf{diag}(\mathbf{X}(\mathrm{u}-)))' d\mathbf{M}(t),\ t \in \mathcal{T}.\end{aligned}$$

Here $\mathbf{N}^{\mathbf{X}}(0) = \mathbf{0}$, the zero vector in $\mathbb{R}^N$.

The following lemma is due to Dufour and Elliott (1999) [14] and gives a compensated version of $\mathbf{N}^{\mathbf{X}}$ under $\mathcal{P}$, which is a martingale associated with $\mathbf{N}^{\mathbf{X}}$. We state the result here without giving the proof.

Lemma 7.4.1 *Let* $\mathbf{A}_0(t) = \mathbf{A}(t) - \mathbf{diag}(\mathbf{a}(t))$, *where* $\mathbf{a}(t) = (a_{11}(t), a_{22}(t), \ldots, a_{NN}(t))' \in \mathbb{R}^N$, *for each* $t \in \mathcal{T}$. *Then the process* $\widetilde{\mathbf{N}}^{\mathbf{X}} = \left\{ \widetilde{\mathbf{N}}^{\mathbf{X}}(t) | t \in \mathcal{T} \right\}$ *defined by putting*

$$\widetilde{\mathbf{N}}^{\mathbf{X}}(t) = \mathbf{N}^{\mathbf{X}}(t) - \int_0^t \mathbf{A}_0(\mathrm{u})\mathbf{X}(\mathrm{u}-)d\mathrm{u}, \quad t \in \mathcal{T} \tag{7.4.17}$$

is an $\mathbb{R}^N$*-valued* $(\mathcal{F}^{\mathbf{X}}, \mathcal{P})$*-martingale.*

Consider the $\mathcal{G}$-adapted process on $(\Omega, \mathcal{F}, \mathcal{P})$ associated with $\mathbf{C}$ defined by setting

$$\Lambda^{\mathbf{C}}(t) = 1 + \int_0^t \Lambda^{\mathbf{C}}(\mathrm{u}-)\left[\mathbf{D}_0^{\mathbf{C}}(\mathrm{u})\mathbf{X}(\mathrm{u}-) - \mathbf{1}\right]' d\widetilde{\mathbf{N}}^{\mathbf{X}}(\mathrm{u}).$$

Here $\mathbf{1} = (1, 1, \ldots, 1)' \in \mathbb{R}^N$.

Then we have the following result.

Lemma 7.4.2 $\Lambda^{\mathbf{C}}$ *is a* $(\mathcal{G}, \mathcal{P})$*-martingale and hence* $\mathrm{E}\left[\Lambda^{\mathbf{C}}(T)\right] = 1$.

Proof This is due to Lemma 7.4.1 and the boundedness of $c_{ij}(t)$, for each $j = 1, 2, \ldots, N$ and each $t \in \mathcal{T}$.

Suppose $\mathcal{K}$ is the space of all families of rate matrices $\mathbf{C}$ with bounded components. Then for each $\mathbf{C} \in \mathcal{K}$, we use $\Lambda^{\mathbf{C}}$ as a density process for a measure change for the chain $\mathbf{X}$.

Consider a $\mathcal{G}$-adapted process $\Lambda^{\theta,\mathbf{C}} := \{\Lambda^{\theta,\mathbf{C}}(t) | t \in \mathcal{T}\}$ on $(\Omega, \mathcal{F}, \mathcal{P})$ defined by

$$\Lambda^{\theta,\mathbf{C}}(t) := \Lambda^{\theta}(t) \cdot \Lambda^{\mathbf{C}}(t), \quad t \in \mathcal{T}.$$

Our assumptions together with Lemma 7.4.2 ensure that $\Lambda^{\theta,\mathbf{C}}$ is a $(\mathcal{G}, \mathcal{P})$-martingale. Consequently, $\mathrm{E}\left[\Lambda^{\theta,\mathbf{C}}(T)\right] = 1$. Then $\Lambda^{\theta,\mathbf{C}}$ can be a density process for a general measure change for the Brownian motion w, the jump process Z and the Markov chain $\mathbf{X}$.

For each pair $(\theta, \mathbf{C}) \in \Theta \times \mathcal{K}$, we define a probability measure $\mathcal{P}^{\theta,\mathbf{C}}$ absolutely continuous with respect to $\mathcal{P}$ on $\mathcal{G}(T)$ as

$$\frac{d\mathcal{P}^{\theta,\mathbf{C}}}{d\mathcal{P}}|_{\mathcal{G}(T)} := \Lambda^{\theta,\mathbf{C}}(T). \tag{7.4.18}$$

Consequently, we can define a family $\mathcal{P}(\Theta \times \mathcal{K})$ of real-world probability measures $\mathcal{P}^{\theta,\mathbf{C}}$ parameterized by $(\theta, \mathbf{C}) \in \Theta \times \mathcal{K}$.

Here the market can choose a real-world probability measure, or generalized 'scenario', from $\mathcal{P}(\Theta)$ by selecting a process $\theta \in \Theta$. So Θ represents the set of admissible strategies, or controls, of the market. By choosing different $\theta \in \Theta$ different probability laws for the price process of the share index and the insurance risk process are obtained. The following theorem gives the probability laws of the Brownian motion w, the random measure γ and the chain $\mathbf{X}$ under the new measure $\mathcal{P}^{\theta,\mathbf{C}}$.

Theorem 7.4.1 *The process defined by*

$$w^{\theta}(t) := w(t) - \int_0^t \theta(\mathrm{u})du, \quad t \in \mathcal{T}$$

is a $(\mathcal{G}, \mathcal{P}^{\theta,\mathbf{C}})$-standard Brownian motion. Furthermore, under $\mathcal{P}^{\theta,\mathbf{C}}$, the random measure γ has the following compensator:

$$\nu^{\theta}_{\mathbf{X}(\mathrm{u}-)}(dz, d\mathrm{u}) = \sum_{i=1}^{N} (1 - \theta(\mathrm{u}-))\lambda_i F_i(dz)d\mathrm{u},$$

and the chain $\mathbf{X}$ has a family of rate matrices $\mathbf{C}$ and can be represented as

$$\mathbf{X}(t) = \mathbf{X}(0) + \int_0^t \mathbf{C}(\mathrm{u})\mathbf{X}(\mathrm{u}-)du + \mathbf{M}^{\mathbf{C}}(t),$$

where $\mathbf{M}^{\mathbf{C}} := \{\mathbf{M}^{\mathbf{C}}(t) | t \in \mathcal{T}\}$ is $(\mathcal{G}, \mathcal{P}^{\theta,\mathbf{C}})$-martingale.

Proof The proof follows from a general Girsanov theorem for jump-diffusion processes and a Girsanov transform for a Markov chain.

The following corollary gives the surplus process under $\mathcal{P}^{\theta,\mathbf{C}}$.

Corollary 7.4.1 *Let $\gamma^{\theta,\mathbf{C}}(dz, dt)$ be a random measure having the compensator $\nu^{\theta}_{\mathbf{X}(t-)}(dz, dt)$ under $\mathcal{P}^{\theta,\mathbf{C}}$. Then, under $\mathcal{P}^{\theta,\mathbf{C}}$ the surplus process of the insurance company evolves over time as*

$$
\begin{aligned}
dV(t) &= [p(t) + r(t)V(t) + \pi(t)(\mu(t) - r(t)) - \sigma(t)\pi(t)\theta(t))]dt \\
&\quad + \sigma(t)\pi(t)w^{\theta}(t) - \int_0^{\infty} z\gamma^{\theta,\mathbf{C}}(dz, dt), \quad t \in \mathcal{T}, \\
V(0) &= v_0.
\end{aligned}
$$

Proof The result follows from Theorem 7.4.1.

Now for each $(\pi, \theta, \mathbf{C}) \in \mathcal{A} \times \Theta \times \mathcal{K}$, we consider a vector-valued, controlled, state process $\{\mathbf{Y}^{\pi,\theta,\mathbf{C}}(t) | t \in \mathcal{T}\}$ defined by

$$
\begin{aligned}
d\mathbf{Y}^{\pi,\theta,\mathbf{C}}(t) &= (dY_0(t), dY_1^{\pi,\theta,\mathbf{C}}(t), d\mathbf{Y}_2^{\mathbf{C}}(t))' = (dt, dV^{\pi,\theta,\mathbf{C}}(t), d\mathbf{X}(t))', \\
\mathbf{Y}^{\pi,\theta,\mathbf{C}}(0) &= \mathbf{y} = (s, y_1, \mathbf{y}_2)',
\end{aligned}
$$

where under the new probability measure $\mathcal{P}^{\theta,\mathbf{C}}$,

$$
\begin{aligned}
dY_0(t) &= dt, \\
dY_1^{\pi,\theta,\mathbf{C}}(t) &= [p(t) + r(t)Y_1^{\pi,\theta,\mathbf{C}}(t) + \pi(t)(\mu(t) - r(t)) - \sigma(t)\pi(t)\theta(t))]dt \\
&\quad + \sigma(t)\pi(t)w^{\theta}(t) - \int_0^{\infty} z\gamma^{\theta,\mathbf{C}}(dz, dt), \\
d\mathbf{Y}_2^{\mathbf{C}}(t) &= \mathbf{C}(t)\mathbf{Y}_2^{\mathbf{C}}(t-)dt + d\mathbf{M}^{\mathbf{C}}(t).
\end{aligned}
\tag{7.4.19}
$$

To simplify the notation, we suppress the superscripts π and θ and write, for each $t \in \mathcal{T}$,

$$
\begin{aligned}
\mathbf{Y}^{\pi,\theta,\mathbf{C}}(t) &:= \mathbf{Y}(t), \\
Y_1^{\pi,\theta,\mathbf{C}}(t) &:= Y_1(t), \\
\mathbf{Y}_2^{\mathbf{C}}(t) &:= \mathbf{Y}_2(t).
\end{aligned}
\tag{7.4.20}
$$

Note that the vector-valued, controlled, state process $\mathbf{Y}$ is Markov with respect to the enlarged filtration $\mathcal{G}$ under $\mathcal{P}$.

We can now formulate the optimal investment problem. Let $U(\cdot) : (0, \infty) \to \mathcal{R}$ denote a strictly increasing and strictly convex utility. Then, conditional on $\mathbf{Y}(0) = \mathbf{y}$, the object of the insurance company is to find a portfolio process $\pi \in \mathcal{A}$ so as to maximize the following minimal expected utility on terminal surplus over the family $\mathcal{P}(\Theta \times \mathcal{K})$:

$$
\inf_{\theta \in \Theta, \mathbf{C} \in \mathcal{K}} \mathrm{E}_{\mathbf{y}}^{\theta,\mathbf{C}}\left[U(Y_1^{\pi}(T))\right].
$$

Here $\mathrm{E}_{\mathbf{y}}^{\theta,\mathbf{C}}$ is the conditional expectation given $\mathbf{Y}(0) = \mathbf{y}$ under $\mathcal{P}^{\theta,\mathbf{C}}$. The minimal expected utility can be interpreted as the expected utility on terminal surplus in the 'worst-case' scenario.

The optimal investment problem of the insurance company can now be described as the following two player, zero-sum, stochastic differential game between the insurance company and the market.

Problem I Consider

$$\Phi(\mathbf{y}) = \sup_{\pi \in \mathcal{A}} \left(\inf_{(\theta, \mathbf{C}) \in \Theta \times \mathcal{K}} \mathbf{E}_{\mathbf{y}}^{\theta, \mathbf{C}} [U(Y_1^{\pi, \theta, \mathbf{C}}(T))] \right) = \mathbf{E}_{\mathbf{y}}^{\theta^*, \mathbf{C}^*} \left[U(Y_1^{\pi^*, \theta^*, \mathbf{C}^*}(T)) \right]. \quad (7.4.21)$$

To solve the related problem I, we need to find the value function $\Phi(\mathbf{y})$ and the optimal strategies $\pi^* \in \mathcal{A}$ and $(\theta^*, \mathbf{C}^*) \in \Theta \times \mathcal{K}$ of the insurance company and the market, respectively.

7.4.2.2 Maximizing the Minimal Survival Probability

In this case, $\mathcal{T} := [0,\infty)$. So the optimal investment problem is one in an infinite-time horizon setting. Firstly, we need to generate a family of real-world probability measures which are absolutely continuous with respect to the reference measure $\mathcal{P}$ in this infinite time horizon setting. We adopt some results for measure changes of Elliott (1982) [19] to define the Radon–Nikodym derivative for the measure changes in the infinite-time horizon setting. Again, we also use the Girsanov theorem for a Markov chain to define a measure change.

Consider a process, denoted as $\bar{\theta} = \{\bar{\theta}(t) | t \in \mathcal{T}\}$, which parameterizes the family of real-world probability measures. Suppose the process $\bar{\theta}$ satisfies the following conditions:

(1) $\bar{\theta}$ is $\bar{\mathcal{G}}$-progressively measurable; and
(2) for each $T < \infty$
$\int_0^T \bar{\theta}^2(t)dt < \infty, \quad \mathcal{P}-\text{a.s.}$

We denote the space of all such processes $\bar{\theta}$ by $\overline{\Theta}$.

Define, for each $\bar{\theta} \in \overline{\Theta}$, a real-valued, $\bar{\mathcal{G}}$-adapted, process $\left\{\overline{\Lambda}^{\bar{\theta}}(t) | t \in \mathcal{T}\right\}$ on $(\Omega, \mathcal{F}, \mathcal{P})$ by setting

$$\overline{\Lambda}^{\bar{\theta}}(t) := \exp\left(- \int_0^t \bar{\theta}(u) dw(u) - \int_0^t \bar{\theta}(u) dw^2(u) - \int_0^t \bar{\theta}^2(u) du \right). \quad (7.4.22)$$

Applying Itô's differentiation rule to $\overline{\Lambda}^{\bar{\theta}}$ gives

$$\begin{aligned} d\overline{\Lambda}^{\bar{\theta}}(t) &= \overline{\Lambda}^{\bar{\theta}}(t)(-\bar{\theta}(t)dw(t) - \bar{\theta}(t)dw^2(t)), \\ \overline{\Lambda}^{\bar{\theta}}(0) &= 1, \quad \mathcal{P}-\text{a.s.}, \end{aligned} \quad (7.4.23)$$

and so $\overline{\Lambda}^{\bar{\theta}}$ is a $(\bar{\mathcal{G}}, \mathcal{P})$-martingale. Suppose $\overline{\Lambda}^{\bar{\theta}}$ is a uniformly integrable positive $(\bar{\mathcal{G}}, \mathcal{P})$-martingale. Then $\overline{\Lambda}^{\theta}(\infty) = \lim_{t \to \infty} \overline{\Lambda}^{\bar{\theta}}(t), \quad \mathcal{P}$–a.s. and

$$\mathrm{E}[\overline{\Lambda}^{\bar{\theta}}(\infty) | \bar{\mathcal{G}}(t)] = \overline{\Lambda}^{\bar{\theta}}(t), \quad \mathcal{P}\text{–a.s.,} \quad \mathrm{t} \in \mathcal{T}. \tag{7.4.24}$$

Hence,

$$\mathrm{E}[\overline{\Lambda}^{\bar{\theta}}(\infty)] = \mathrm{E}[\overline{\Lambda}^{\bar{\theta}}(0)] = 1. \tag{7.4.25}$$

Again, we consider the density process $\Lambda^{\mathbf{C}}$ for the measure change of the Markov chain $\mathbf{X}$ defined in the last subsection. We suppose that $\Lambda^{\mathbf{C}}$ is a uniformly integrable positive $(\bar{\mathcal{G}}, \mathcal{P})$-martingale. Then $\Lambda^{\mathbf{C}}(\infty) = \lim_{t \to \infty} \Lambda^{\mathbf{C}}(t), \quad \mathcal{P}$–a.s. and

$$\mathrm{E}[\overline{\Lambda}^{\mathbf{C}}(\infty) | \bar{\mathcal{G}}(t)] = \overline{\Lambda}^{\bar{\theta}}(t), \quad \mathcal{P}\text{–a.s.,} \quad \mathrm{t} \in \mathcal{T}. \tag{7.4.26}$$

Consequently,

$$\mathrm{E}[\Lambda^{\mathbf{C}}(\infty)] = \mathrm{E}[\Lambda^{\mathbf{C}}(0)] = 1. \tag{7.4.27}$$

For each $(\bar{\theta}, \mathbf{C}) \in \overline{\Theta} \times \mathcal{K}$ a new (real-world) probability measure $\mathcal{P}^{\bar{\theta}, \mathbf{C}}$ equivalent to $\mathcal{P}$ can be defined one $(\Omega, \bar{\mathcal{G}}(\infty))$ by putting

$$\frac{d\mathcal{P}^{\bar{\theta}, \mathbf{C}}}{d\mathcal{P}}\bigg|_{\bar{\mathcal{G}}(\infty)} = \overline{\Lambda}^{\bar{\theta}}(\infty) \cdot \Lambda^{\mathbf{C}}(\infty). \tag{7.4.28}$$

A family $\mathcal{P}(\overline{\Theta} \times \mathcal{K})$ of real-world probability measures parameterized by $(\bar{\theta}, \mathbf{C}) \in \overline{\Theta} \times \mathcal{K}$ is then generated. Here, $\overline{\Theta} \times \mathcal{K}$ represents the set of admissible strategies, or controls, of the market.

The following theorem then gives the surplus process $\bar{V}$ and the semi-martingale dynamics of the Markov chain $\mathbf{X}$ under the new probability measure $\mathcal{P}^{\bar{\theta}, \mathbf{C}}$.

Theorem 7.4.2 *Define the processes* $w^{\theta,z} = \{w^{\theta,z}(t) | t \in \mathcal{T}\}$ *and* $w^{\theta} = \{w^{\theta}(t) | t \in \mathcal{T}\}$ *by*

$$w^{\theta,z} := w^z(t) - \int_0^t \theta(\mathrm{u}) du,$$
$$w^{\theta}(t) := w(t) - \int_0^t \theta(\mathrm{u}) du.$$

Then under $\mathcal{P}^{\bar{\theta}, \mathbf{C}}$, $w^{\theta,z}$, w^{θ} are standard Brownian motions. Furthermore, under $\mathcal{P}^{\bar{\theta}, \mathbf{C}}$, the surplus process of the insurance company evolves over time as

$$d\bar{V}(t) = [p(t) + r(t)\bar{V}(t) + \bar{\pi}(t)(u(t) - r(t)) - \theta(t)(\sigma(t)\bar{\pi}(t) - \sigma_z(t))]dt \\ + \sigma(t)\bar{\pi}(t)dw^{\theta}(t) - \sigma_z(t)dw^{\theta,z}(t),$$

and the Markov chain $\mathbf{X}$ has the following semi-martingale dynamics:

$$\mathbf{X}(t) = \mathbf{X}(0) + \int_0^t \mathbf{C}(\mathrm{u})\mathbf{X}(\mathrm{u})d\mathrm{u} + \mathbf{M}^{\mathbf{C}}(t).$$

Similarly to the previous subsection, for each $(\bar{\pi}, \bar{\theta}, \mathbf{C}) \in \bar{\mathcal{A}} \times \overline{\Theta} \times \mathcal{K}$, we consider here a vector-valued, controlled, state process $\left\{\overline{\mathbf{Y}}^{\bar{\pi},\bar{\theta},\mathbf{C}}(t) | t \in \mathcal{T}\right\}$ defined by

$$d\overline{\mathbf{Y}}^{\bar{\pi},\bar{\theta},\mathbf{C}}(t) = (d\bar{Y}_1^{\bar{\pi},\bar{\theta},\mathbf{C}}(t), d\overline{\mathbf{Y}}_2^{\mathbf{C}}(t))' = (d\bar{V}^{\bar{\pi},\bar{\theta},\mathbf{C}}(t), d\mathbf{X}(t))'\overline{\Lambda}, \\ \overline{\mathbf{Y}}^{\bar{\pi},\bar{\theta},\mathbf{C}}(0) = \bar{\mathbf{y}} = (\bar{y}_1, \mathbf{y}_2)'$$

where under $\mathcal{P}$,

$$d\bar{Y}_1^{\bar{\pi},\bar{\theta},\mathbf{C}}(t) = [p(t) + r(t)\bar{Y}_1^{\bar{\pi},\bar{\theta},\mathbf{C}}(t) + \bar{\pi}(t)(\mu(t) - r(t)) - \theta(t)(\sigma(t)\bar{\pi}(t) - \sigma_z(t))]dt \\ + \sigma(t)\bar{\pi}(t)w^{\theta}(t) - \sigma_z(t)dw^{\theta,z}, \\ d\mathbf{Y}_2^{\mathbf{C}}(t) = \mathbf{C}(t)\mathbf{Y}_2^{\mathbf{C}}(t-)dt + d\mathbf{M}^{\mathbf{C}}(t). \tag{7.4.29}$$

To simplify the notation, we write

$$\overline{\mathbf{Y}}^{\bar{\pi},\bar{\theta},\mathbf{C}}(t) := \bar{\mathbf{Y}}(t), \\ \bar{Y}_1^{\bar{\pi},\bar{\theta},\mathbf{C}}(t) := \bar{Y}_1(t), \\ \overline{\mathbf{Y}}_2^{\mathbf{C}}(t) := \overline{\mathbf{Y}}_2(t). \tag{7.4.30}$$

It is obvious that the vector-valued, controlled, state process $\overline{\mathbf{Y}}$ is Markov with respect to the enlarged filtration $\bar{\mathcal{G}}$ under $\mathcal{P}$.

We suppose that ruin occurs when the surplus of the insurance company goes zero. For each portfolio process $\bar{\pi} \in \bar{\mathcal{A}}$, let $\tau^{\bar{\pi}} : =\tau^{\bar{\pi}}(\cdot) : \Omega \to [0, \infty]$ be the first time that the surplus process $\bar{V}_1^{\bar{\pi}}$ of the insurance company reaches zero, that is

$$\tau^{\bar{\pi}} = \inf\{t \in \mathcal{T} | \bar{V}_1^{\bar{\pi}}(t) = 0\}. \tag{7.4.31}$$

Here $\tau^{\bar{\pi}}$ represents the ruin time of the company. Indeed, it is a predictable stopping time.

Conditional on $\overline{\mathbf{Y}}(0) = \bar{\mathbf{y}}$, the object of the insurance company is to select an investment strategy $\bar{\pi} \in \bar{\mathcal{A}}$ which maximizes the following minimal survival probability over the family $\mathcal{P}(\overline{\Theta} \times \mathcal{K})$ of real-world probability measures:

$$\inf_{(\bar{\theta},\mathbf{C}) \in \overline{\Theta} \times \mathcal{K}} \mathcal{P}^{\bar{\theta},\mathbf{C}}(\tau^{\bar{\pi}} = \infty | \overline{\mathbf{Y}}(0) = \bar{\mathbf{y}}) = \inf_{(\bar{\theta},\mathbf{C}) \in \overline{\Theta} \times \mathcal{K}} \mathrm{E}_{\bar{\mathbf{y}}}^{\bar{\theta},\mathbf{C}}[I\{\tau^{\bar{\pi}} = \infty\}]. \tag{7.4.32}$$

Here $\mathrm{E}_{\bar{\mathbf{y}}}^{\bar{\theta},\mathbf{C}}$ is the conditional expectation given $\overline{\mathbf{Y}}(0) = \bar{\mathbf{y}}$. The minimal survival probability can be interpreted as the survival probability in the 'worst-case' scenario.

The optimal investment problem of the company can then be formulated as the following stochastic differential game.

Problem II Consider

$$\overline{\Phi}(\bar{\mathbf{y}}) = \sup_{\bar{\pi} \in \bar{\mathcal{A}}} \left(\inf_{(\bar{\theta},\mathbf{C}) \in \overline{\Theta} \times \mathcal{K}} \mathrm{E}_{\bar{\mathbf{y}}}^{\bar{\theta},\mathbf{C}}[I\{\tau^{\bar{\pi}} = \infty\}] \right) = \mathbf{E}_{\bar{\mathbf{y}}}^{\bar{\theta}^*,\mathbf{C}^*}\left[I\{\tau^{\bar{\pi}^*} = \infty\}\right]. \tag{7.4.33}$$

To solve the problem, we must find $\bar{\pi}^* \in \bar{\mathcal{A}}$, $(\bar{\theta}^*, \mathbf{C}^*) \in \overline{\Theta} \times \mathcal{K}$ and $\overline{\Phi}(\bar{\mathbf{y}})$.

7.4.3 Solution to Optimal Investment Problem I

In this section, we adopt the HJB dynamic programming approach to solve the optimal investment problem I. We first give a verification theorem for the HJB solution to problem I. Then closed-form expressions for the optimal strategies of the game are derived under some assumptions.

Firstly, we specify the relationship between the control processes of the game and the information structure. Note that the controlled state process $\mathbf{Y}$ is adapted to the enlarged filtration $\mathcal{G}$ and that it is also Markov with respect to $\mathcal{G}$.

Let $\mathcal{O} = (0, T) \times (0, \infty)$ so that $\mathcal{O} \times \mathcal{E}$ is our solvency region. Suppose K_1 denotes the subset of $\mathbb{R}$ such that $\pi(t) \in K_1$, for each $t \in \mathcal{T}$. Similarly, let K_2 be a subset in $\mathbb{R}$ such that $\theta(t) \in K_2$, for each $t \in \mathcal{T}$. Suppose $\mathbf{K}_3$ is a subset of $\mathbb{R}^N \otimes \mathbb{R}^N$ such that $\mathbf{C}(t) \in \mathbf{K}_3, \mathcal{P}$–a.s. for each $t \in \mathcal{T}$, where $\mathbb{R}^N \otimes \mathbb{R}^N$ is the space of $N \times N$ matrices. Here we assume that $\mathbf{K}_3$ is a rectangular region so that for each $i,j = 1, 2, \ldots, N$, $t \in \mathcal{T}$.

$$c_{ij}(t) \in \left[c^l(i,j), c^u(i,j)\right].$$

for some given constants $c^l(i,j)$ and $c^u(i,j)$ satisfying the following conditions:

(1) for each $k = l, u$ and $i \neq j, c^k(i,j) \in [0, \infty)$;
(2) $\sum_{i=1}^N c^k(i,j) = 0$, $c^k(i,j) \leq 0$;
(3) $c^l(i,j) < c^u(i,j)$, (i.e. the interval $\left[c^l(i,j), c^u(i,j)\right]$ is not degenerate).

We suppose that $\pi(t) = \tilde{\pi}(\mathbf{Y}(t))$, $\theta(t) = \tilde{\theta}(\mathbf{Y}(t))$, $\mathbf{C}(t) = \widetilde{\mathbf{C}}(\mathbf{Y}(t))$, for some functions $\tilde{\pi} : \mathcal{O} \times \mathcal{E} \to K_1$, $\tilde{\theta} : \mathcal{O} \times \mathcal{E} \to K_2$, $\widetilde{\mathbf{C}} : \mathcal{O} \times \mathcal{E} \to \mathbf{K}_3$.

In what follows, with a slight abuse of notation, we do not distinguish between π and $\tilde{\pi}$, and between $\mathbf{C}$ and $\widetilde{\mathbf{C}}$, and between θ and $\tilde{\theta}$. Then we can identify the control $\pi(t)$, $\mathbf{C}(t)$ and $\theta(t)$ with deterministic functions $\pi(\mathbf{y})$, $\mathbf{C}(\mathbf{Y})$, $\theta(\mathbf{y})$, respectively, for each $\mathbf{y} \in \mathcal{O} \times \mathcal{E}$. These are called feedback controls.

Since both the state process and the control are Markov, the dynamic programming principle can be applied. We first present a verification theorem for the HJB solution to the stochastic differential game corresponding to problem I.

Let $\mathcal{H}$ be the space of functions $h(\cdot,\cdot,\cdot) : \mathcal{T} \times \mathbb{R}^+ \times \mathcal{E} \to \mathbb{R}$ such that for each $\mathbf{e}_i \in \mathcal{E}$, $h(\cdot,\cdot,\mathbf{e}_i) \in \mathcal{C}^{1,2}(\mathcal{T} \times \mathbb{R}^+)$. Write

$$\mathbf{h}(s, y_1) := (h(s, y_1, \mathbf{e}_1), h(s, y_1, \mathbf{e}_2), \ldots, h(s, y_1, \mathbf{e}_N))' \in \mathbb{R}^N.$$

Then for each $(\theta, \mathbf{C}, \pi) \in \Theta \times \mathcal{K} \times \mathcal{A}$, the generator of the process $\mathbf{Y}$ under the new measure $\mathcal{P}^{\theta,\mathbf{C}}$ is a partial differential operator $\mathcal{L}^{\theta,\mathbf{C},\pi}$ acting on $\mathcal{H}$:

$$\begin{aligned}\mathcal{L}^{\theta,\mathbf{C},\pi}[h(\mathbf{y})] &= \frac{\partial h}{\partial s} + [p(s) + r(s)h + \pi(\mu(t) - r(t) - \sigma(t)\pi\theta)]\frac{\partial h}{\partial y_1} \\ &\quad + \frac{1}{2}\sigma^2(s)\pi_2^2\frac{\partial^2 h}{\partial y_1^2} + \int_0^\infty (h(s, y_1 - z, \mathbf{y}_2) - h(s, y_1, \mathbf{y}_2)) \\ &\quad \times (1-\theta)\lambda(s)F_{\mathbf{y}_{2-}}(dz) + \langle \mathbf{h}(s, y_1), \mathbf{C}'\mathbf{y}_{2-}\rangle,\end{aligned} \tag{7.4.34}$$

where $\mathbf{y}_{2-} := \mathbf{Y}_2(s-)$. Then we have the following lemma, which will be used to prove the verification theorem.

Lemma 7.4.3 *Suppose for each $\mathbf{e}_i \in \mathcal{E}$, $h(\cdot,\cdot,\mathbf{e}_i) \in \mathcal{C}^{1,2}(\mathcal{T} \times \mathbb{R}^+)$. Let τ be an (optional) stopping time such $\tau < \infty$ s. Assume, further, that for each $\mathbf{e}_i \in \mathcal{E}$ and $(\theta, \mathbf{C}, \pi) \in \Theta \times \mathcal{K} \times \mathcal{A}$, $h(s, y_1, \mathbf{e}_i)$ and $\mathcal{L}^{\theta,\mathbf{C},\pi}[h(s, y_1, \mathbf{e}_i)]$ are bounded on $(s, y_1) \in \mathcal{T} \times \mathbb{R}^+$. Then*

$$\mathbf{E}_{\mathbf{y}}^{\theta,\mathbf{C}}[h(\mathbf{Y}(\tau))] = h(\mathbf{y}) + \mathbf{E}_{\mathbf{y}}^{\theta,\mathbf{C}}\left(\int_0^\tau \mathcal{L}^{\theta,\mathbf{C},\pi}[h(s, Y_1(s), \mathbf{Y}_2(s))]ds\right). \tag{7.4.35}$$

Proof Applying Itô's differentiation rule for semimartingales to $h(\mathbf{Y}(t))$, then, combine (7.4.34) and integral on $[0, \tau]$, we can get Lemma 7.4.3.

We present a verification theorem for the HJB solution to problem I in the following proposition.

Theorem 7.4.3 *Let $\bar{\mathcal{O}}$ be the closure of $\mathcal{O}$. Suppose there exists a function ϕ and a Markov control $(\theta^*, \mathbf{C}^*, \pi^*) \in \Theta \times \mathcal{K} \times \mathcal{A}$, such that, for each $\mathbf{e}_i \in \mathcal{E}$, $\phi(\cdot,\cdot,\mathbf{e}_i) \in \mathcal{C}^2(\mathcal{O}) \cap \mathcal{C}(\bar{\mathcal{O}})$,*

(1) $\mathcal{L}^{\theta,\mathbf{C},\pi^*(\mathbf{y})}[\phi(s, y_1, \mathbf{y}_2)] \geq 0$, *for all* $(\theta, \mathbf{C}) \in \Theta \times \mathcal{K}$ *and* $(s, y_1, \mathbf{y}_2) \in \mathcal{O} \times \mathcal{E}$;

(2) $\mathcal{L}^{\theta^*(\mathbf{y}),\mathbf{C}^*(\mathbf{y}),\pi}[\phi(s,y_1,\mathbf{y}_2)] \leq 0$, *for all* $\pi \in \mathcal{A}$ *and* $(s,y_1,\mathbf{y}_2) \in \mathcal{O} \times \mathcal{E}$;
(3) $\mathcal{L}^{\theta^*(\mathbf{y}),\mathbf{C}^*(\mathbf{y}),\pi^*(\mathbf{y})}[\phi(s,y_1,\mathbf{y}_2)] = 0$, *for all* $(s,y_1,\mathbf{y}_2) \in \mathcal{O} \times \mathcal{E}$;
(4) *for all* $(\theta,\mathbf{C},\pi) \in \Theta \times \mathcal{K} \times \mathcal{A}$, $\lim_{t\to T^-} \phi(\mathbf{Y}^{\theta,\mathbf{C},\pi}(t)) = U(Y_1^{\theta,\mathbf{C},\pi}(T))$;
(5) *Let* $\bar{\mathcal{M}}$ *denote the set of stopping times* $\tau := \tau(\omega) \leq T$, *for all* $\omega \in \Omega$, *the family* $\{\phi(\mathbf{Y}^{\theta,\mathbf{C},\pi}(\tau))\}_{\tau\in\mathcal{K}}$ *is uniformly integrable.*

Define, for each $(s,y_1,\mathbf{y}_2) \in \mathcal{O} \times \mathcal{E}$ and $(\theta,\pi) \in \Theta \times \mathcal{A}$

$$J^{\theta,\pi}(\mathbf{y}) = \mathrm{E}_{\mathbf{y}}^{\theta,\mathbf{C}}\left[U(Y^{\theta,\mathbf{C},\pi}(T))\right]. \tag{7.4.36}$$

Then

$$\begin{aligned}\phi(y) &= \Phi(y)\\ &= \inf_{(\theta,\mathbf{C})\in\Theta\times\mathcal{K}}\left(\sup_{\pi\in\mathcal{A}} \mathrm{E}_{\mathbf{y}}^{\theta,\mathbf{C}}\left[U(Y_1^{\theta,\mathbf{C},\pi}(T))\right]\right)\\ &= \sup_{\pi\in\mathcal{A}}\left(\inf_{(\theta,\mathbf{C})\in\Theta\times\mathcal{K}} \mathrm{E}_{\mathbf{y}}^{\theta,\mathbf{C}}\left[U(Y_1^{\theta,\mathbf{C},\pi}(T))\right]\right)\\ &= \sup_{\pi\in\mathcal{A}} J^{\theta^*,\mathbf{C}^*,\pi}(\mathbf{y}) = \inf_{(\theta,\mathbf{C})\in\Theta\times\mathcal{K}} J^{\theta,\mathbf{C},\pi^*}(\mathbf{y})\\ &= J^{\theta^*,\mathbf{C}^*,\pi^*}(\mathbf{y}).\end{aligned} \tag{7.4.37}$$

and $(\theta^*,\mathbf{C}^*,\pi^*)$ is an optimal Markov control.

In the sequel, we derive an explicit solution to problem I when the insurer has an exponential utility. We also need to assume that the interest rate $r(t) = 0$, for each $t \in \mathcal{T}$, in the derivation of the explicit solution.

Let $U(\cdot) : (0,\infty) \to \mathbb{R}$ denote an exponential utility function defined by

$$U(x) = -\mathrm{e}^{\alpha x}, \tag{7.4.38}$$

where is α positive constant, which represents the coefficient of absolute risk aversion; that is

$$\alpha = -\frac{U_{xx}(x)}{U_x(x)}.$$

with U_x and U_{xx} representing the first and second derivatives of U with respect to x.

We try the following parametric form for the value function:

$$\Phi(s,y_1,\mathbf{y}_2) = \mathrm{e}^{-\alpha y_1} g(s,\mathbf{y}_2), \tag{7.4.39}$$

where $g : \mathcal{T} \times \mathcal{E} \to \mathbb{R}$ is a function such that, for each $\mathbf{y}_2 \in \mathcal{E}$, $g(T,y_2) = 1$.

We can re-state conditions 1–3 of Theorem 7.4.3 as

$$\inf_{(\theta,\mathbf{C})\in\Theta\times\mathcal{K}} \mathcal{L}^{\theta,\mathbf{C},\pi^*}[\Phi(\mathbf{y})] = \mathcal{L}^{\theta^*,\mathbf{C}^*,\pi^*}[\Phi(\mathbf{y})] = 0 \tag{7.4.40}$$

and

$$\sup_{\pi\in\mathcal{A}} \mathcal{L}^{\theta^*,\mathbf{C}^*,\pi}[\Phi(\mathbf{y})] = \mathcal{L}^{\theta^*,\mathbf{C}^*,\pi^*}[\Phi(\mathbf{y})] = 0. \tag{7.4.40}$$

When $r(t) = 0$, the partial differential operator $\mathcal{L}^{\theta,\mathbf{C},\pi}[\Phi(\mathbf{y})]$ becomes

$$\begin{aligned}\mathcal{L}^{\theta,\mathbf{C},\pi}[\Phi(\mathbf{y})] =& \frac{\partial\Phi}{\partial s} + [p(s) + \pi\mu(t) - \sigma(t)\pi\theta]\frac{\partial\Phi}{\partial y_1} + \frac{1}{2}\sigma^2(s)\pi^2\frac{\partial^2\Phi}{\partial y_1^2} \\ &+ \int_0^\infty (\Phi(s, y_1 - z, \mathbf{y}_2) - \Phi(s, y_1, \mathbf{y}_2))(1-\theta)\lambda(s)F_{\mathbf{y}_{2-}}(dz) + \langle \mathbf{\Phi}(s, y_1), \mathbf{C}'\mathbf{y}_{2-}\rangle,\end{aligned}$$

where $\mathbf{\Phi}(s, y_1) = (\Phi(s, y_1, \mathbf{e}_1), \Phi(s, y_1, \mathbf{e}_1), \ldots, \Phi(s, y_1, \mathbf{e}_N))' \in \mathbb{R}^N$.

Let $g(s) := (g(s, \mathbf{e}_1), g(s, \mathbf{e}_1), \ldots, g(s, \mathbf{e}_N))' \in \mathbb{R}^N$, for each $s \in \mathcal{T}$. Then it is not difficult to see that

$$\begin{aligned}\mathcal{L}^{\theta,\mathbf{C},\pi}[\Phi(\mathbf{y})] =& \mathrm{e}^{-\alpha y_1}\Bigg[\frac{dg(s, \mathbf{y}_2)}{ds} - (\alpha p(s) + \alpha\pi\mu(s) - \alpha\sigma(s)\pi\theta - \frac{1}{2}\alpha^2\sigma^2(s)\pi^2 \\ &+ (\theta - 1)\lambda(s)\int_0^\infty (\mathrm{e}^{\alpha z} - 1)F_{\mathbf{y}_{2-}}(dz))g(s, \mathbf{y}_{2-}) + \langle \mathbf{g}(s), \mathbf{C}'\mathbf{y}_{2-}\rangle\Bigg].\end{aligned}$$

So the first-order condition for a value π^* to maximize $\mathcal{L}^{\theta,\mathbf{C},\pi}[\Phi(\mathbf{y})]$ over all $\pi \in \mathcal{A}$ is

$$\alpha\mathrm{e}^{-\alpha y_1} g(s, \mathbf{y}_{2-})(-\mu(s) + \sigma^2(s)\pi^*\alpha + \theta^*\sigma(s)) = 0. \tag{7.4.41}$$

Similarly, the first-order condition for a value θ^* to maximize $\mathcal{L}^{\theta,\mathbf{C},\pi}[\Phi(\mathbf{y})]$ over all $\pi \in \mathcal{A}$ is

$$\mathrm{e}^{-\alpha y_1} g(s, \mathbf{y}_{2-})\left(\sigma(s)\pi^*\alpha + \lambda(s)\int_0^\infty (1 - \mathrm{e}^{\alpha z})F_{\mathbf{y}_{2-}}(dz)\right) = 0. \tag{7.4.42}$$

Therefore, we obtain the following closed-form solutions of the optimal strategies θ^* and π^*:

$$\begin{aligned}\theta^*(s, \mathbf{X}(s)) &= \sum_{i=1}^N (\frac{\mu_i - \sigma_i^2\pi^*(s, \mathbf{e}_i)\alpha}{\sigma_i})\langle \mathbf{X}(s), \mathbf{e}_i\rangle, \\ \pi^*(s, \mathbf{X}(s)) &= \sum_{i=1}^N \left(\frac{\lambda_i \int_0^\infty (\mathrm{e}^{\alpha z} - 1)F_{\mathbf{e}_i}(dz)}{\sigma_i\alpha}\right)\langle \mathbf{X}(s), \mathbf{e}_i\rangle.\end{aligned} \tag{7.4.43}$$

We now need to determine a value $\mathbf{C}^*$ to minimize $\mathcal{L}^{\theta,\mathbf{C},\pi}[\Phi(\mathbf{y})]$ over all $C < \mathcal{K}$. To simplify the notation, we write $\Phi_i = \Phi(s, y_1, \mathbf{e}_i)$, for $i = 1, 2, \ldots, N$ and $\mathbf{\Phi} = \mathbf{\Phi}(s, y_1)$. Firstly, we note that the partial differential $\mathcal{L}^{\theta^*,\mathbf{C},\pi^*}$ acting on $\Phi(s, y_1, \mathrm{y}_2)$ is equivalent to the following system of partial differential operators $\mathcal{L}_j^{\theta^*,\mathbf{C},\pi^*}$ acting on Φ_j.

$$\mathcal{L}_j^{\theta^*,\mathbf{C},\pi^*}[\Phi_j] = \frac{\partial \Phi_j}{\partial s} + [p(s) + \pi\mu_j - \sigma(t)\pi^*\theta^*]\frac{\partial \Phi}{\partial y_1} + \frac{1}{2}\sigma_j^2(\pi^*)^2\frac{\partial^2 \Phi}{\partial y_1^2}$$
$$+ \int_0^\infty (\Phi(s, y_1 - z, \mathbf{y}_2) - \Phi(s, y_1, \mathbf{y}_2))(1 - \theta^*)\lambda_j(s)F_{\mathbf{e}_j}(dz) + \sum_{i=1}^{N} \Phi(i)c_{ij}(t),'$$
$$j = 1, 2, \ldots, N.$$

Since the only part of $\mathcal{L}_j^{\theta^*,\mathbf{C},\pi^*}[\Phi_j]$ that depends on $\mathbf{C}$ is the sum $\sum_{i=1}^{N} \Phi(i)c_{ij}(t)$, the minimization of $\mathcal{L}_j^{\theta^*,\mathbf{C},\pi^*}[\Phi_j]$ with respect to $\mathbf{C}$ is equivalent to the following system of N linear programming problems:

$$\min_{c_{ij}(t), c_{ij}(t), \ldots, c_{Nj}(t)} \sum_{i=1}^{N} \Phi(i)c_{ij}(t), \quad j = 1, 2, \ldots, N.$$

subject to the linear constraints

$$\sum_{i=1}^{N} c_{ij}(t) = 0,$$

and the 'interval' constraints

$$c_{ij}(t) \in \left[c^l(i,j), c^u(i,j)\right], \quad i,j = 1, 2, \ldots, N.$$

Note that the linear constraints come from the property of rate matrices and the 'interval' constraints are due to the rectangularity of $\mathbf{K}_3$.

When the Markov chain $\mathbf{X}$ has two states, we can determine the optimal strategy $\mathbf{C}^*$ explicitly by solving the following pair of linear programming problems:

$$\max_{c_{11}(t), c_{21}(t)} [\Phi_1 c_{11}(t) + \Phi_2 c_{21}(t)],$$
$$s.t.$$
$$c_{11}(t) + c_{21}(t) = 0, \quad c_{11}(t) \in \left[c^l(1,1), c^u(1,1)\right], \quad t \in \mathcal{T}.$$

and

$$\min_{c_{11}(t),c_{21}(t)} [\Phi_1 c_{12}(t) + \Phi_2 c_{22}(t)],$$
$$s.t.$$
$$c_{12}(t) + c_{22}(t) = 0, \quad c_{12}(t) \in \left[c^l(1,2), c^u(1,2)\right], \quad t \in \mathcal{T}.$$

The solutions of the pair of linear programming problems are

$$c^*_{11}(t) = c^l(1,1) I_{\{\Phi_1 - \Phi_2 > 0\}} + c^u(1,1) I_{\{\Phi_1 - \Phi_2 < 0\}},$$
$$c^*_{21}(t) = -c^*_{11}(t), \quad t \in \mathcal{T},$$

and

$$c^*_{12}(t) = c^l(1,2) I_{\{\Phi_1 - \Phi_2 > 0\}} + c^u(1,2) I_{\{\Phi_1 - \Phi_2 < 0\}},$$
$$c^*_{22}(t) = -c^*_{12}(t), \quad t \in \mathcal{T}.$$

The explicit form of the optimal strategy $\mathbf{C}^*$ is, therefore $\mathbf{C}^*(t) = \left[c^*_{ij}(t)\right]_{i,j=1,2,\ldots,N}$.

Note that $\mathcal{L}^{\theta^*,\mathbf{C}^*,\pi^*}[\phi(\mathbf{y})] = 0$. Then, it is not difficult to see that $g(s, \mathbf{e}_i)$, satisfy the following system of coupled, first-order, linear ordinary differential equations (ODEs):

$$\frac{dg(s,\mathbf{e}_i)}{ds} - \left(\alpha p_i + \alpha\mu_i\pi^*_i - \frac{1}{2}\alpha^2\sigma_i^2(\pi^*_i)^2 - \alpha\theta^*_i\sigma_i\pi^*_i \right.$$
$$\left. + (\theta^*_i - 1)\lambda_i \int_0^\infty (\mathrm{e}^{\alpha z} - 1) F_{\mathbf{e}_i}(dz) + c^*_{ii}\right) g(s,\mathbf{e}_i) + \sum_{j=1, j\neq i}^{N} g(s,\mathbf{e}_i) c^*_{ji} = 0, \tag{7.4.44}$$

with terminal condition $g(T, \mathbf{e}_i)$. Consequently, this is a system of backward ODEs.

For each $i = 1, 2, \ldots, N$, define

$$K_i(\pi^*_i, \theta^*_i, c^*_{ii}) := \alpha p_i + \alpha\mu_i\pi^*_i - \frac{1}{2}\alpha^2\sigma_i^2(\pi^*_i)^2 - \alpha\theta^*_i\sigma_i\pi^*_i$$
$$+ (\theta^*_i - 1)\lambda_i \int_0^\infty (\mathrm{e}^{\alpha z} - 1) F_{\mathbf{e}_i}(dz) + c^*_{ii}.$$

Then a version of the variation-of-constant formula gives

$$g(s,\mathbf{e}_i) = \mathrm{e}^{K_i(\pi^*_i,\theta^*_i,c^*_{ii})(T-s)} + \int_s^T \mathrm{e}^{K_i(\pi^*_i,\theta^*_i,c^*_{ii})(\mathrm{u}-s)} \left(\sum_{j=1,j\neq i}^{N} c^*_{ji} g(\mathrm{u},\mathbf{e}_j)\right) d\mathrm{u}. \tag{7.4.45}$$

7.4.4 Solution to Problem II

In this section, we first present a general stochastic differential game which includes problem II as a particular case. Then we give a verification theorem for the HJB solution to the general game, and semi-analytical solutions to the optimal strategies to problem II are derived under some assumptions.

Let $\mathcal{S} := (l, u)$, for each $l, u \in \mathbb{R}$ with $l<u$, so that $\mathcal{S} \times \mathcal{E}$ is the solvency region. Write $\bar{\mathcal{S}}$ for the closure of $\mathcal{S}$ and $\tau_{\mathcal{S}}^{\bar{\pi}} = \inf\{t \geq 0 | \bar{Y}^{\bar{\pi}}(t) \notin \mathcal{S}\}$ for the bankruptcy time associated with the portfolio process $\bar{\pi} \in \bar{\mathcal{A}}$. To simplify the notation, we suppress the subscript $\bar{\pi}$ and write $\tau_{\mathcal{S}} = \tau_{\mathcal{S}}^{\bar{\pi}}$. Let $\beta(x)$ be a non-negative continuous function, $F(x)$ a real-valued bounded continuous function, and $H(x)$ a real-valued function. we define $(\bar{\theta}, \mathbf{C}, \bar{\pi}) \in \overline{\Theta} \times \mathcal{K} \times \bar{\mathcal{A}}$ by

$$\bar{J}^{\bar{\theta},\mathbf{C},\bar{\pi}}(\bar{\mathbf{y}}) := \mathrm{E}_{\bar{\mathbf{y}}}^{\bar{\theta},\mathbf{C}}\left[\int_0^{\tau_{\mathcal{S}}} F(\overline{\mathbf{Y}}^{\bar{\pi}}(t))\mathrm{e}^{-\int_0^t \beta(\overline{\mathbf{Y}}^{\bar{\pi}}(s))ds} dt + H(\overline{\mathbf{Y}}^{\bar{\pi}}(\tau_{\mathcal{S}}))\mathrm{e}^{-\int_0^{\tau_{\mathcal{S}}} \beta(\overline{\mathbf{Y}}^{\bar{\pi}}(t))dt}\right]. \tag{7.4.46}$$

Then the general stochastic differential game can be formulated as

$$\overline{\Phi}(\mathbf{y}) = \sup_{\bar{\pi}\in\bar{\mathcal{A}}}\left(\inf_{(\bar{\theta},\mathbf{C})\in\overline{\Theta}\times\mathcal{K}} \bar{J}^{\bar{\theta},\mathbf{C},\bar{\pi}}(\bar{\mathbf{y}})\right). \tag{7.4.47}$$

Again, we consider Markov controls and assume that $\bar{\theta}(t) := \bar{\theta}(\overline{\mathbf{Y}}(t))$, $\mathbf{C}(t) = \mathbf{C}(\overline{\mathbf{Y}}(t))$ and $\bar{\pi}(t) = \bar{\pi}(\overline{\mathbf{Y}}(t))$. We further suppose that the optimal strategies $\bar{\theta}^*(\bar{\mathbf{y}})$, $\mathbf{C}^*(\bar{\mathbf{y}})$ and $\bar{\pi}^*(\bar{\mathbf{y}})$ exist and that $|\overline{\Phi}(\bar{\mathbf{y}})| < \infty$.

Let $\bar{\mathcal{H}}$ denote the space of function $\bar{h}(\cdot,\cdot) : \mathcal{S} \times \mathcal{E} \to (-\infty, \infty)$ such that for each $\mathbf{e}_i \in \mathcal{E}$, $\bar{h}(\cdot, \mathbf{e}_i) \in \mathcal{C}^2(\mathcal{S})$. Write $\bar{\mathbf{h}}(\bar{y}_1) := (\bar{h}(y_1, \mathbf{e}_1), \bar{h}(y_1, \mathbf{e}_2), \ldots, \bar{h}(y_1, \mathbf{e}_N))' \in \mathbb{R}^N$. Then, for each $(\bar{\theta}, \mathbf{C}, \bar{\pi}) \in \overline{\Theta} \times \mathcal{K} \times \bar{\mathcal{A}}$, we consider the following partial differential operator $\bar{\mathcal{L}}^{\bar{\theta},\mathbf{C},\bar{\pi}}$ acting on $\bar{\mathcal{H}}$:

$$\begin{aligned}\bar{\mathcal{L}}^{\bar{\theta},\mathbf{C},\bar{\pi}}[\bar{h}(\bar{\mathbf{y}})] &= [p(t) + r(t)\bar{h} + \bar{\pi}(\mu(t) - r(t)) - \bar{\theta}(\sigma(t)\bar{\pi} - \sigma_z(t))]\frac{\partial \bar{h}}{\partial \bar{y}_1} \\ &\quad + \frac{1}{2}(\sigma^2(t)\bar{\pi}^2 + 2\rho(t)\sigma(t)\sigma_z(t)\bar{\pi} + \sigma_z^2(t))\frac{\partial^2 \bar{h}}{\partial \bar{y}_1^2} - \beta(\bar{\mathbf{y}})\bar{h} + \langle \bar{\mathbf{h}}(y_1), \mathbf{C}'\mathbf{y}_2\rangle,\end{aligned} \tag{7.4.48}$$

Then we give the following lemma.

Lemma 7.4.4 *Suppose $\bar{h}(\bar{\mathbf{y}}) \in \bar{\mathcal{H}}$. Let τ be an (optional) stopping time such that $\tau = \tau(\omega) < \infty, \mathcal{P}-$a.s. Assume further that $\bar{h}(\bar{\mathbf{Y}}(s))$ and $\bar{\mathcal{L}}^{\bar{\theta},\mathbf{C},\bar{\pi}}[\bar{h}(\bar{\mathbf{Y}}(s))]$ are bounded on $s \in [0, \tau]$, then*

$$\mathrm{E}_{\bar{\mathbf{y}}}^{\bar{\theta},\mathbf{C}}[\mathrm{e}^{-\int_0^t \beta(\bar{\mathbf{Y}}^{\bar{\pi}}(s))ds}\bar{h}(\bar{\mathbf{Y}}^{\bar{\pi}}(\tau))] = \bar{h}(\bar{\mathbf{y}}) + \mathrm{E}_{\bar{\mathbf{y}}}^{\bar{\theta},\mathbf{C}}\left[\int_0^{\tau} \mathrm{e}^{-\int_0^s \beta(\bar{\mathbf{Y}}^{\bar{\pi}}(u))du}\bar{\mathcal{L}}^{\bar{\theta},\mathbf{C},\bar{\pi}}[\bar{h}(\bar{\mathbf{Y}}^{\bar{\pi}}(s))]ds\right] \tag{7.4.49}$$

The proof of Lemma 7.4.4 resembles that of Lemma 7.4.1, so we do not repeat it here.

The following theorem gives a verification theorem for the HJB solution for the general game.

Theorem 7.4.3 *Suppose there exists a function $\bar{\phi}(\cdot,\cdot) : \mathcal{S} \times \mathcal{E} \to (-\infty, \infty)$ such that for each $\mathbf{e}_i \in \mathcal{E}$, $\bar{\phi} \in (.,\mathbf{e}_i)\mathcal{C}^2(\mathcal{S}) \cap \mathcal{C}(\bar{\mathcal{S}})$, and a Markov control $(\bar{\theta}^*(\bar{\mathbf{y}}), \mathbf{C}^*(\bar{\mathbf{y}}), \bar{\pi}^*(\bar{\mathbf{y}})) \in \overline{\Theta} \times \mathcal{K} \times \bar{\mathcal{A}}$ such that*

(1) $\bar{\mathcal{L}}^{\bar{\theta},\mathbf{C},\bar{\pi}^*(\bar{\mathbf{y}})}[\bar{\phi}(\bar{\mathbf{y}})] + F(\bar{\mathbf{y}}) \geq 0$, *for all* $(\bar{\theta}, \mathbf{C}) \in \overline{\Theta} \times \mathcal{K}$, $\bar{\mathbf{y}} \in \mathcal{S} \times \mathcal{E}$;

(2) $\bar{\mathcal{L}}^{\bar{\theta}^*(\bar{\mathbf{y}}),\mathbf{C}^*(\bar{\mathbf{y}}),\bar{\pi}}[\bar{\phi}(\bar{\mathbf{y}})] + F(\bar{\mathbf{y}}) \leq 0$, *for all* $\bar{\pi} \in \bar{\mathcal{A}}$, $\bar{\mathbf{y}} \in \mathcal{S} \times \mathcal{E}$;

(3) $\bar{\mathcal{L}}^{\bar{\theta}^*(\bar{\mathbf{y}}),\mathbf{C}^*(\bar{\mathbf{y}}),\bar{\pi}^*(\bar{\mathbf{y}})}[\phi(\mathbf{y})] + F(\bar{\mathbf{y}}) = 0$, *for all* $\bar{\pi} \in \bar{\mathcal{A}}$, $\bar{\mathbf{y}} \in \mathcal{S} \times \mathcal{E}$;

(4) on $\{\tau_{\mathcal{S}} < \infty\}$, $\bar{\mathbf{Y}}^{\bar{\theta},\mathbf{C},\bar{\pi}}(\tau_{\mathcal{S}}) \in \partial\mathcal{S} \times \mathcal{E}$, *where $\partial\mathcal{S}$ is the boundary of $\mathcal{S}$, and*

$$\lim_{t \to \tau_{\mathcal{S}}^-} \bar{\phi}(\bar{\mathbf{Y}}^{\bar{\theta},\mathbf{C},\bar{\pi}}(t)) = H(\bar{\mathbf{Y}}^{\bar{\theta},\mathbf{C},\bar{\pi}}(\tau_{\mathcal{S}}))I\{\tau_{\mathcal{S}} < \infty\};$$

(5) *let $\bar{\mathcal{M}}$ denote the space of $\bar{\mathcal{G}}$-stopping times $\tau(\omega) \leq \tau_{\mathcal{S}}(\omega)$, for all $\omega \in \Omega$. The family $\left\{\bar{\phi}(\bar{\mathbf{Y}}^{\bar{\theta},\mathbf{C},\bar{\pi}}(\tau))\right\}_{\tau \in \bar{\mathcal{K}}}$ is uniformly integrable, for all $\bar{\mathbf{y}} \in \mathcal{S} \times \mathcal{E}$ and $(\bar{\theta}, \mathbf{C}, \bar{\pi}) \in \overline{\Theta} \times \mathcal{K} \times \bar{\mathcal{A}}$.*
Then, for all $\bar{\mathbf{y}} \in \mathcal{S} \times \mathcal{E}$

$$\bar{\phi}(y) = \overline{\Phi}(y) = \sup_{\bar{\pi} \in \bar{\mathcal{A}}} J^{\bar{\theta}^*,\mathbf{C}^*,\bar{\pi}}(\bar{\mathbf{y}}) = \inf_{(\bar{\theta},\mathbf{C}) \in \overrightarrow{\Theta} \times \mathcal{K}} J^{\bar{\theta},\mathbf{C},\bar{\pi}^*}(\bar{\mathbf{y}}) = J^{\bar{\theta}^*,\mathbf{C}^*,\bar{\pi}^*}(\bar{\mathbf{y}}). \tag{7.4.50}$$

and $(\bar{\theta}^*, \mathbf{C}^*, \bar{\pi}^*)$ is an optimal Markov control.

The proof of Theorem 7.4.3 resembles that of Theorem 7.4.2.

In what follows, we derive the solution to Problem II. In this case, $l = 0$, $u = \infty$, $F(x) = 0$ and $\beta(x) = 0$. We also assume that $r(t) = 0$, for each $t \in \mathcal{T}$. In this case, the partial differential operator becomes

$$\bar{\mathcal{L}}^{\bar{\theta},\mathbf{C},\bar{\pi}}[\bar{h}(\bar{\mathbf{y}})] = \left[p(t) + \bar{\pi}\mu(t) - \bar{\theta}(\sigma(t)\bar{\pi} - \sigma_z(t))\right]\frac{\partial \bar{h}}{\partial \bar{y}_1} + \frac{1}{2}(\sigma^2(t)\bar{\pi}^2 + 2\rho(t)\sigma(t)\sigma_z(t)\bar{\pi} + \sigma_z^2(t))\frac{\partial^2 \bar{h}}{\partial \bar{y}_1^2} - \beta(\bar{\mathbf{y}})\bar{h} + \langle \bar{\mathbf{h}}(y_1), \mathbf{C}'\mathbf{y}_2\rangle. \tag{7.4.51}$$

Firstly, the value function satisfies the following boundary conditions:

$$\overline{\Phi}(\bar{y}_1, \bar{\mathbf{y}}_2) = 0, \quad \bar{y}_1 = 0, \tag{7.4.52}$$

$$\overline{\Phi}(\bar{y}_1, \bar{\mathbf{y}}_2) = 1, \quad \bar{y}_1 \to \infty. \tag{7.4.53}$$

We try a solution of the following form:

$$\Phi(\bar{\mathbf{y}}) = \left(\kappa(\bar{\mathbf{y}}_2) - \frac{\gamma(\bar{\mathbf{y}}_2)}{\eta(\bar{\mathbf{y}}_2)}e^{-\eta(\bar{\mathbf{y}}_2)\bar{y}_1}\right), \tag{7.4.54}$$

where $\eta(\cdot) : \mathcal{E} \to \mathbb{R}^+$, $\gamma(\cdot) : \mathcal{E} \to \mathbb{R}^+$ and $\kappa(\cdot) : \mathcal{E} \to \mathbb{R}^+$ are some functions.
The boundary conditions imply that

$$\overline{\Phi}(\bar{\mathbf{y}}) = (1 - \mathrm{e}^{-\eta(\bar{\mathbf{y}}_2)\bar{y}_1}). \tag{7.4.55}$$

Note that $F(x) = 0$. Conditions (1)–(3) of Theorem 7.4.3 read

$$\inf_{(\bar{\theta},C)\in\overline{\Theta}\times\mathcal{K}} \bar{\mathcal{L}}^{\bar{\theta},\mathbf{C},\bar{\pi}^*}\left[\overline{\Phi}(\bar{\mathbf{y}})\right] = \bar{\mathcal{L}}^{\bar{\theta}^*,\mathbf{C}^*,\bar{\pi}^*}\left[\overline{\Phi}(\bar{\mathbf{y}})\right] = 0. \tag{7.4.56}$$

$$\sup_{\bar{\pi}\in\bar{\mathcal{A}}} \bar{\mathcal{L}}^{\bar{\theta}^*,\mathbf{C}^*,\bar{\pi}}\left[\overline{\Phi}(\bar{\mathbf{y}})\right] = \bar{\mathcal{L}}^{\bar{\theta}^*,\mathbf{C}^*,\bar{\pi}^*}\left[\overline{\Phi}(\bar{\mathbf{y}})\right] = 0. \tag{7.4.57}$$

Note that

$$\bar{\mathcal{L}}^{\bar{\theta},C,\bar{\pi}}[\overline{\Phi}(\bar{\mathbf{y}})] = \eta(\bar{\mathbf{y}}_2)\mathrm{e}^{-\eta(\bar{\mathbf{y}}_2)\bar{y}_1}\left[p(s) + \bar{\pi}\mu(s) - \bar{\theta}(\sigma(s)\bar{\pi} - \sigma_z(s)) - \frac{1}{2}(\sigma^2(s)\bar{\pi}^2 + 2\rho(s)\sigma(s)\sigma_z(s)\bar{\pi} + \sigma_z^2(s))\eta(\bar{\mathbf{y}}_2)\right] + \langle\overline{\Phi}(y_1), \mathbf{C}'\mathbf{y}_2\rangle, \tag{7.4.58}$$

where $\overline{\mathbf{\Phi}}(\bar{y}_1) := (\overline{\Phi}(\bar{y}_1, \mathbf{e}_1), \overline{\Phi}(\bar{y}_1, \mathbf{e}_2), \ldots, \overline{\Phi}(\bar{y}_1, \mathbf{e}_N))' \in \mathbb{R}^N$.
The first-order condition for a minimum point $\bar{\theta}^*$ of $\bar{\mathcal{L}}^{\bar{\theta},C,\bar{\pi}}\left[\overline{\Phi}(\bar{y})\right]$ gives

$$\bar{\pi}^*(\bar{\mathbf{y}}_2) = \frac{\sigma_z(s)}{\sigma(s)} = \sum_{i=1}^{N}\left(\frac{\sigma_{zi}}{\sigma_i}\right)\langle\bar{\mathbf{y}}_2, \mathbf{e}_i\rangle. \tag{7.4.59}$$

Similarly, the first-order condition for a maximum point $\bar{\pi}^*$ of $\bar{\mathcal{L}}^{\bar{\theta},C,\bar{\pi}}\left[\overline{\Phi}(\bar{y})\right]$ gives

$$\bar{\theta}^*(\bar{\mathbf{y}}_2) = \frac{\mu(s) - (\rho(s) - 1)\sigma(s)\sigma_z(s)\eta(\bar{\mathbf{y}}_2)}{\sigma(s)}. \tag{7.4.60}$$

As in the previous section, we have to determine a value $\mathbf{C}^* \in \mathcal{K}$ to minimize $\bar{\mathcal{L}}^{\bar{\theta}^*,C,\bar{\pi}^*}$ over all $\mathbf{C} \in \mathcal{K}$. For each $i = 1, 2, \ldots, N$, let $\overline{\Phi}_i = \overline{\Phi}(\bar{y}_1, \mathbf{e}_i)$. Write $\overline{\mathbf{\Phi}} := \overline{\mathbf{\Phi}}(\bar{y}_1)$. Then the partial differential operator $\bar{\mathcal{L}}^{\bar{\theta}^*,C,\bar{\pi}^*}$ acting on $\overline{\Phi}(\bar{y}_1)$ is equivalent to the following system of partial differential operators $\bar{\mathcal{L}}_j^{\bar{\theta}^*,\mathbf{C},\bar{\pi}^*}$ acting on $\overline{\Phi}_j$

$$\begin{aligned}\bar{\mathcal{L}}_j^{\bar{\theta}^*,\mathbf{C},\bar{\pi}^*}[\overline{\Phi}_j] &= \eta(\mathbf{e}_j)\mathrm{e}^{-\eta(\mathbf{e}_j)\bar{y}_1}\left[p_j + \bar{\pi}^*\mu_j - \bar{\theta}^*(\sigma_j\bar{\pi}^* - \sigma_{zj})\right.\\ &\quad \left. - \frac{1}{2}(\sigma_j^2(\bar{\pi}^*)^2 - 2\rho_j\sigma_j\sigma_{zj}\bar{\pi}^* + \sigma_{zj}^2)\eta(\mathbf{e}_j)\right] + \sum_{i=1}^{N}\overline{\Phi}(i)c_{ij}(t), \quad j = 1, 2, \ldots, N.\end{aligned} \tag{7.4.61}$$

Again, the minimization problem is equivalent to the following N linear programming problems:

$$\min_{c_{1j}(t), c_{2j}(t), \ldots, c_{Nj}(t)} \sum_{i=1}^{N} \Phi(i)c_{ij}(t), \quad j = 1, 2, \ldots, N,$$

subject to the linear constraints

$$\sum_{i=1}^{N} c_{ij}(t) = 0,$$

and the 'interval' constraints

$$c_{ij}(t) \in \left[c^l(i,j), c^u(i,j)\right], \quad i,j = 1, 2, \ldots, N.$$

When the Markov chain $\mathbf{X}$ has two states the solutions of the system of linear programming problems are

$$\begin{aligned}c_{11}^*(t) &= c^l(1,1)I_{\{\Phi_1 - \Phi_2 > 0\}} + c^u(1,1)I_{\{\Phi_1 - \Phi_2 < 0\}},\\ c_{21}^*(t) &= -c_{11}^*(t), \quad t \in \mathcal{T},\end{aligned}$$

and

$$\begin{aligned}c_{12}^*(t) &= c^l(1,2)I_{\{\Phi_1 - \Phi_2 > 0\}} + c^u(1,2)I_{\{\Phi_1 - \Phi_2 < 0\}},\\ c_{22}^*(t) &= -c_{12}^*(t), \quad t \in \mathcal{T}.\end{aligned}$$

Consequently, the optimal strategy $\mathbf{C}^*$ is, therefore, $\mathbf{C}^*(t) = \left[c_{ij}^*(t)\right]_{i,j=1,2,\ldots,N}$.

Note that $\bar{\mathcal{L}}_j^{\bar{\theta}^*,\mathbf{C},\bar{\pi}^*}[\overline{\Phi}(\bar{\mathbf{y}})] = 0$. Then $\eta(\mathbf{e}_j)$, $j = 1,2,\ldots,N$, satisfy the following system of N nonlinear equations:

$$\eta(\mathbf{e}_j)\mathrm{e}^{-\eta(\mathbf{e}_j)\bar{y}_1}\left[p_j + \bar{\pi}^*\mu_j - \bar{\theta}^*\left(\sigma_j\bar{\pi}^* - \sigma_{zj}\right) - \frac{1}{2}\left(\sigma_j^2(\bar{\pi}^*)^2 - 2\rho_j\sigma_j\sigma_{zj}\bar{\pi}^*\right)\eta(\mathbf{e}_j)\right]$$
$$+ \sum_{i=1}^{N}\mathrm{e}^{-\eta(\mathbf{e}_i)\bar{y}_1}c_{ij}^*(t) = 0, \quad j = 1,2,\ldots,N.$$

7.5 Summary

We using a game theoretic model under linear Markov jump systems which obtained from the previous chapters. Firstly, a risk minimization problem of portfolio is considered in Markovian regime switching. And then, we obtained the equilibrium solution of European option pricing problem under Markovian regime switching. We adopted a robust approach to describe model uncertainty and formulated the optimal investment problems as two-player, zero-sum, stochastic differential games between the market and the insurance company, and deduced the optimal strategy of closed game.

References

1. Asmussen, S. (1989). Exponential families generated by phase-type distributions and other Markov lifetime. *Scandinavian Journal of Statistics, 16*(4), 319–334.
2. Yeung, D. W. K., & Petrosyan, L. A. (2006). *Cooperative stochastic differential games*. New York: Springer.
3. Haller, H., & Lagunoff, R. (2000). Genericity and Markovian behavior in stochastic games. *Econometrica, 68*(6), 1231–1248.
4. Heath, D., Delbaen, F., Eber, J. M., et al. (1999). Coherent measures of risk. *Mathematical Finance, 9*(3), 203–228.
5. Asmussen, S. (1989). Risk theory in a Markovian environment. *Scandinavian Actuarial Journal, 1989*(2), 69–100.
6. Browne, S. (1995). Optimal investment policies for a firm with a random risk process: exponential utility and minimizing the probability of ruin. *Mathematics of Operations Research, 20*(4), 937–958.
7. Browne, S. (1997). Survival and growth with a liability: Optimal portfolio strategies in continuous time. *Mathematics of Operations Research, 22*(2), 468–493.
8. Browne, S. (1999). Beating a moving target: Optimal portfolio strategies for outperforming a stochastic benchmark. *Finance and Stochastics, 3*(3), 275–294.
9. Buffington, J., & Elliott, R. J. (2002). American options with regime switching. *International Journal of Theoretical and Applied Finance, 5*(5), 497–514.
10. Cairns, A. J. G. (2000). A discussion of parameter and model uncertainty in insurance. *Insurance: Mathematics and Economics, 27*(3), 313–330.

11. Cvitanić, J., & Karatzas, I. (1999). On dynamic measures of risk. *Finance and Stochastics, 3* (4), 451–482.
12. Daykin, C. D., Pentikainen, T., & Pesonen, M. (1993). *Practical risk theory for actuaries*. CRC Press.
13. Risk, M. (1996). *Quantitative strategies research notes*. Goldman Sachs.
14. Dufour, F., & Elliott, R. J. (1999). Filtering with discrete state observations. *Applied Mathematics and Optimization, 40*(2), 259–272.
15. Elliott, R. (1976). The existence of value in stochastic differential games. *SIAM Journal on Control and Optimization, 14*(1), 85–94.
16. Elliott, R. J. (1977). The existence of optimal controls and saddle points in stochastic differential games. In *Proceedings of the Workshop on Differential Games, Enschede. Lecture Notes in Control and Information Sciences* (Vol. 3, pp. 136–142).
17. Mataramvura, S., & Øksendal, B. (2008). Risk minimizing portfolios and HJBI equations for stochastic differential games. *Stochastics An International Journal of Probability and Stochastic Processes, 80*(4), 317–337.
18. Elliott, R. J. (1977). The optimal control of a stochastic system. *SIAM Journal on Control and Optimization, 15*(5), 756–778.
19. Elliott, R. J. (1982). *Stochastic calculus and applications*. Springer.
20. Föllmer, H., & Schied, A. (2002). Convex measures of risk and trading constraints. *Finance and Stochastics, 6*(4), 429–447.
21. Huang, L., & Mao, X. (2011). Stability of singular stochastic systems with Markovian switching. *IEEE Transactions on Automatic Control, 56*(2), 424–429.
22. Elliott, R. J. (1990). Filtering and control for point process observations. *Recent Advances in Stochastic Calculus* 1–27.
23. Friedman, M. (2007). *Price theory*. Transaction Publishers.
24. Stigler, G. J. (1994). *Production and distribution theories*. Transaction Publishers.
25. Lucas Jr., R. E. (1978). Asset prices in an exchange economy. *Econometrica: Journal of the Econometric Society* 1429–1445.
26. Øksendal, B., & Sulem, A. (2008). A game theoretic approach to martingale measures in incomplete markets. *Surveys of Applied and Industrial Mathematics, 15*, 18–24.
27. Harrison, J. M., & Kreps, D. M. (1979). Martingales and arbitrage in multiperiod securities markets. *Journal of Economic theory, 20*(3), 381–408.
28. Harrison, J. M., & Pliska, S. R. (1981). Martingales and stochastic integrals in the theory of continuous trading. *Stochastic Processes and Their Applications, 11*(3), 215–260.
29. Harrison, J. M., & Pliska, S. R. (1983). A stochastic calculus model of continuous trading: Complete markets. *Stochastic Processes and Their Applications, 15*(3), 313–316.
30. Dybvig, P. H., & Ross, S. A. (1987). Arbitrage. *The New Palgrave: A Dictionary of Economics, 1*, 100–106.
31. Back, K., & Pliska, S. R. (1991). On the fundamental theorem of asset pricing with an infinite state space. *Journal of Mathematical Economics, 20*(1), 1–18.
32. Schachermayer, W. (1992). A Hilbert space proof of the fundamental theorem of asset pricing in finite discrete time. *Insurance: Mathematics and Economics, 11*(4), 249–257.
33. Delbaen, F., & Schachermayer, W. (1994). A general version of the fundamental theorem of asset pricing. *Mathematische Annalen, 300*(1), 463–520.
34. Elliott, R. J., Chan, L., & Siu, T. K. (2005). Option pricing and Esscher transform under regime switching. *Annals of Finance, 1*(4), 423–432.
35. Gerber, H. U., & Shiu, E. S. W. (1994). Option pricing by Esscher transforms. *Transactions of Society of Actuaries, 46*(3), 99–191.
36. Bühlmann, H. (1980). An economic premium principle. *Astin Bulletin, 11*(01), 52–60.
37. Bühlmann, H. (1984). The general economic premium principle. *Astin Bulletin, 14*(01), 13–21.
38. Elliott, R. J., Chan, L., & Siu, T. K. (2005). Option pricing and Esscher transform under regime switching. *Annals of Finance, 1*(4), 423–432.
39. Elliott, R. J., Aggoun, L., & Moore, J. B. (2008). *Hidden Markov models: Estimation and control*. Springer Science & Business Media.

FSC
www.fsc.org
MIX
Papier aus verantwortungsvollen Quellen
Paper from responsible sources
FSC® C105338

If you have any concerns about our products,
you can contact us on
ProductSafety@springernature.com

In case Publisher is established outside the EU,
the EU authorized representative is:
Springer Nature Customer Service Center GmbH
Europaplatz 3, 69115 Heidelberg, Germany

Printed by Libri Plureos GmbH
in Hamburg, Germany